T0234356

# Synthesis Lectures on Mechanical Engineering

This series publishes short books in mechanical engineering (ME), the engineering branch that combines engineering, physics and mathematics principles with materials science to design, analyze, manufacture, and maintain mechanical systems. It involves the production and usage of heat and mechanical power for the design, production and operation of machines and tools. This series publishes within all areas of ME and follows the ASME technical division categories.

Andrew M. Brown

# Structural Dynamics of Liquid Rocket Engines

## A Holistic Approach

Test firing of NASA/MSFC Fastrac rocket engine

 Springer

Andrew M. Brown
NASA Marshall Space Flight Center
Huntsville, AL, USA

ISSN 2573-3168           ISSN 2573-3176   (electronic)
Synthesis Lectures on Mechanical Engineering
ISBN 978-3-031-18209-9        ISBN 978-3-031-18207-5   (eBook)
https://doi.org/10.1007/978-3-031-18207-5

This Springer imprint is published by the registered company Springer Nature Switzerland AG
The registered company address is: Gewerbestrasse 11, 6330 Cham, Switzerland

# Acknowledgements

A number of colleagues have had a significant impact on my career in structural dynamics and in the focus on rocket engines. Ward Winer of Georgia Tech inspired me to pursue engineering in the first place, and Larry Kiefling, my first team lead at NASA, opened the door of this fascinating specialty; he didn't speak out much, but when he did, everyone listened. I'd also like to thank John Townsend, Jeff Peck, Eric Christensen, and the late John Admire at NASA, Gary Davis and Ray Shirr at Rocketdyne, and my colleagues in the GUIde Consortium on Turbine Aeromechanics, especially Rakesh Shrivastava, Matt Montgomery, and Bob Kielb for happily and thoroughly answering my many questions. Gary Genge, Joe Ruf, and Andy Mulder of NASA substantially contributed to my understanding of liquid rocket engines, and Tim Wray, Jen DeLessio, also NASA, and Roland Szabo of Rocketdyne have teamed with me to keep it interesting in recent years. I'd also like to thank Tim for reviewing this entire text as well and Paul Petralia of Morgan and Claypool, a subsidiary of Springer, who encouraged me to pursue this project. Finally, I'd like to acknowledge my late wife, Cheryl, who was always supportive of my work and proud of any accomplishments I achieved.

# Contents

# About the Author

**Andrew M. Brown** Ph. D, is an Aerospace Engineer in the NASA-MSFC/Propulsion Structures & Dynamic Analysis Branch. He joined MSFC in 1986, and has worked mainly on rocket engine dynamics and loads research, development, and production analysis, but also has performed hypersonic debris impact analysis in support of the Columbia Accident Investigation. He has authored or co-authored 11 journal papers and 29 conference papers covering topics ranging from probabilistic design and loads combination methods to techniques for calculating turbine blade forced response in the presence of asymmetric flow. Dr. Brown received a BS in Mechanical Engineering from Duke University and a Masters and Ph.D. in ME from Georgia Tech, and is an Associate Fellow of the American Institute of Astronautics and Aeronautics.

# The Critical Role of Structural Dynamics in the Design, Analysis, Testing, and Operation of Liquid Rocket Engines

**1**

## 1.1 Introduction

The design, analysis, test, and operation of rocket engines powered by liquid propellants such as liquid hydrogen, liquid oxygen, methane, and kerosene is truly a mind-boggling task, and there really is a basis for the term "rocket scientist" as a synonym for excellence. As with all technologies, today's engines come from a steady succession of increasingly complex systems, with each version implementing new breakthroughs. An engineer coming into the workforce today could not possibly have come up with the incredibly convoluted (almost "Rube-Goldberg-like") input–output paths in modern engines where every conceivable watt of power is harvested for propulsion, but they can certainly hope to contribute to the steady evolution necessary to meet each generation's needs.

Along with these increasingly complex machines the need has arisen for specialized engineers with extensive knowledge focused in one of several disciplines. These include fluid dynamics, thermal analysis, combustion dynamics, acoustic analysis, fluid testing, structural testing, rotordynamics, stress analysis, and structural dynamics (SD). Generalists able to pull these various disciplines together are also critical, but the amount of detail in each specialty is far too overwhelming for even a true "renaissance" engineer to be able to master it all. In many ways, this field is like medicine, where specialties within specialties exist (e.g., gynecological oncology). Of course, there is also the continuing ego-battle between which discipline is more difficult, rocket science (really rocket engineering) or brain surgery!

The structural dynamics discipline (not to be confused with rotordynamics, a related but quite separate specialty) plays a vital role in a rocket engine development program. Although it is perhaps not as critical to the design of every single component as stress analysis is, the neglect of proper SD analysis can lead to catastrophic failures. In addition, the definition of somewhat-realistic engine system SD loads at the outset of a design

© The Author(s), under exclusive license to Springer Nature Switzerland AG 2022
A. M. Brown, *Structural Dynamics of Liquid Rocket Engines*, Synthesis Lectures on Mechanical Engineering, https://doi.org/10.1007/978-3-031-18207-5_1

program will have far-reaching effects on the ultimate cost and weight of the engine, which can be show-stoppers.

However, after working in this field for 35 years, I realized that there was very little documentation on basic procedures necessary for even standard, always-required SD analysis of these machines, much less techniques to address more complicated problems. In addition, although there are numerous textbooks on general structural dynamics, there are no publicly available texts on this specialized application of SD, and most structural dynamics engineers in the aerospace industry learn loads methodologies developed for launch vehicles, which are quite different from the techniques necessary for rocket engines. The result is a lack of experienced engineers familiar with these techniques, and a lack of training resources for both that experienced group and for new engineers coming out of college, who only occasionally have even a basic course in mechanical vibrations. I therefore believe that a textbook focused on this topic will be extremely useful for the space industry, both government and commercial.

The book is intended to be a holistic presentation for both engineers new to structural dynamics and experienced structural dynamics engineers who have had experience with launch vehicle loads and vibro-acoustics. The initial third of the book will be a "just what you need to know" summary of basic concepts in vibrations and structural dynamics necessary to do the work, and the rest of the book will focus on specific tasks and problems. The first section will be fairly equation heavy, while the rest of the book will be more text and figure-oriented, with some attention paid to the terminology that is so critical for understanding and communication. Some of the topics examined include the Campbell and SAFE Diagrams for resonance identification in turbomachinery, the complications of component analysis in the pump side due to a host of complicating factors such as acoustic/structure interaction, the "side-loads" fluid/structure interaction problem in over-expanded rocket nozzles and competing methods for generating overall engine system interface loads. The role of modal test for verification will also be discussed. Several topics will be introduced but not elaborated upon in detail; references for further investigation by the reader are listed, though. I will include many specific examples from my experience for illustration. The text will be the first book focused on this indispensable aspect of liquid rocket engine design, and I believe it will fulfill a critical need in the industry and therefore help enable the continued exploration of space.

## 1.2    Overview of Rocket Engine Principles and Design

Although a deep understanding of liquid rocket engine (LRE) operation and design is not required to perform SD analysis, some grasp of the basic principles is necessary, so a brief overview will be presented here. Figure 1.1 shows a schematic of a launch vehicle and engine. The non-payload sections of the vehicle consist largely of storage tanks for the liquid fuel, such as liquid hydrogen, kerosene, or methane, and the liquid

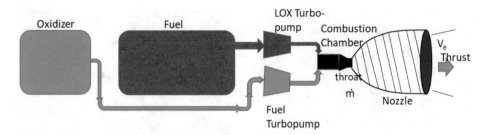

**Fig. 1.1**   Schematic of liquid fueled rocket

oxidizer (oxygen). These tanks are only strong enough to allow pressurization of these propellants to about a hundred psi to avoid excessive weight. To raise the pressure of the propellant high enough for the engine combustion process (approximately 3000 psi for high performance turbomachinery) turbopumps are used. Combustion consists essentially of a spark plug igniting the fuel/oxidizer mixture, and the hot gas then directed through a converging/diverging nozzle which accelerates the exhaust to a supersonic speed. As the supersonic gas then expands in the nozzle, it gains velocity until the exit plane, and the thrust is simply

$$\text{Thrust} = \text{F} = \dot{m} V_e.$$

where $\dot{m}$ is the mass flow rate from the combustion chamber through the throat of the nozzle, and $V_e$ is the exit velocity of the hot gas out of the nozzle.

The design of the engine is dependent on many factors, including thrust, specific impulse $I_{sp}$, which is a measure of rocket engine efficiency defined by the thrust integrated over time per unit propellant weight, and total engine weight. The essential parameters of the engine that affect SD analysis, pressure, temperature, and turbopump speed, are conveniently included in the engine balance (Fig. 1.2). These parameters are outputs of the requirements, and in general cannot be pre-established or set. This makes solving SD problems much more difficult, as a problematic speed, for example, cannot be simply set higher or lower.

Two critical aspects of the SD analysis that cannot be ignored are the thermal conditions and properties and the mean stress state of the component under consideration. Not only does the thermal condition significantly influence the Young's Modulus, but large thermal gradients, which can be particularly severe during engine start-up and shut-down, can introduce large mean stress, which is also critical. The mean stress combines with the alternating stress in a Goodman Diagram for high cycle fatigue or even ultimate failure capability, and also will non-trivially change the component stiffness and therefore the structural dynamics (as we will see in Chap. 2).

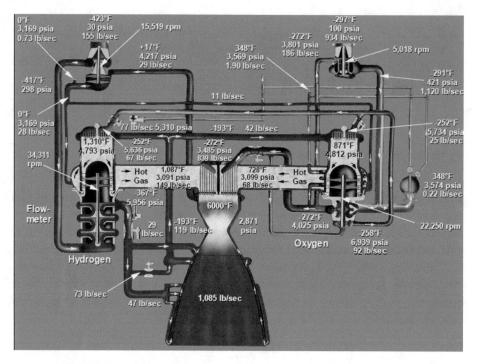

**Fig. 1.2** Power balance for SSME [1]

## 1.3     Failures Due to Structural Dynamics in LRE's

One of the primary motivations for performing accurate SD analysis is to prevent failure. Although documentation of many of these failures have not been publicly released, some data from older failures is available. Cikanek in 1987 stated that "most SSME (Space Shuttle Main Engine) failures were a result of design deficiencies stemming from inadequate definition of dynamic loads" [2]. One of these was the complete burn-through in 1981 of 149 liquid oxidizer injectors (Fig. 1.3), caused by high cycle fatigue cracking resulting from the design being insufficient to withstand huge random loads caused by combustion and flow-induced vortex shedding of the cylindrical injectors [3].

Another major failure of the SSME occurred during a test firing in 1985 in which the manifold carrying heated fuel from the nozzle to the main combustion chamber failed due to high-cycle-fatigue (HCF), resulting in the engine being "severed from the test stand" and winding up in the test-stand spillway (Fig. 1.4). Although the exact cause was never identified, HCF is almost always a result of either resonance or poorly understood dynamic loads. The video of this test and explosion, unfortunately no longer easily available, is spectacular.

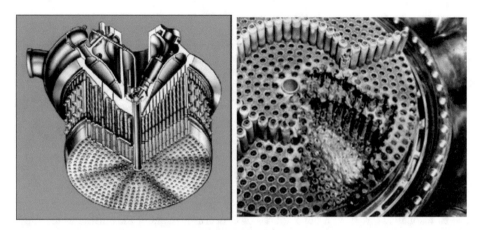

**Fig. 1.3**   SSME main injector failure due to inadequate design margin to dynamic loads

**Fig. 1.4**   SSME lying at bottom of flame trench after manifold HCF failure

Similarly, the RS-68 engine design circa 2001 was re-worked after a turbine blade failure due to resonance was discovered during post-test inspection [4], and several other resonance issues were of concern during the development program. Cracking of the fuel feed line flow-liner in the Space Shuttle Main Propulsion System (frequently considered part of the engine system, Fig. 1.5), which forced a grounding of the fleet for 9 months in 2003, was also discovered to be high-cycle fatigue caused by excitation of a resonance of the structure [5]. These issues are not confined to rocket engines; turbine blade cracking

**Fig. 1.5**  Space shuttle feedline flowliner cracking

due to resonance has been a huge source of concern throughout the entire turbine industry (power generation, jet engines, etc.), and significant resources have been devoted to mitigating these problems [6].

## References

1. Zoladz TF, Lunde K, Mitchell W, SSME Investment in Turbomachinery inducer-Impeller Design Tools, Methodology and Test, 57th JANNAF Joint Propulsion Meeting, 2010
2. Cikanek H, Characteristics of Space Shuttle Main Engine Failures, AIAA Joint Propulsion Conference, 1987, San Diego, CA
3. Goetz O, Monk J, Combustion Device Failures During Space Shuttle Main Engine Development, 5th International Symposium on Liquid Space Propulsion Long Life Combustion Devices Technology, October 27–30, 2003, Chattanooga, Tennessee
4. Ray J, Heart of Boeing's Delta 4 rocket put to the test, Spaceflight Now, May 7, 2001
5. Melcher JC, Rigby, DA, Analysis and Repair of Cracks in the Space Shuttle Main Propulsion System Propellant Feedlines, 39th AIAA Joint Propulsion Conference and Exhibit, 20–23 July 2003, Huntsville, Alabama
6. Srinivasan AV, Flutter and resonant vibration characteristics of engine blades. ASME Journal of Engineering for Gas Turbines and Power. 119 (4): 741–775.

# Structural Dynamics Theory

**2**

## 2.1 Math Review, Units

Although it is assumed that the reader has at least a bachelor's degree knowledge of engineering mathematics, we will briefly review a few mathematical formulas as well as pertinent units in this section that are well worth emphasizing. We first define the coordinate systems used in this text. Referring to Fig. 2.1, the location of a point is defined by capital $X$, $Y$, and $Z$, while the displacements of that point are defined by $u$, $v$, and $w$. While critical for a clear description of structural dynamic models, engineering documentation in the workplace frequently ignores this distinction, where no differentiation is made between location and displacement.

On a related note, we will also use the common shorthand for velocity and acceleration

$$\dot{u} \equiv \frac{\partial u}{\partial t}, \quad \ddot{u} \equiv \frac{\partial^2 u}{\partial t^2}.$$

Although the more rigorous definition of these variables as partials is shown here, in structural dynamic practice there is generally no distinction necessary from the total derivative, which includes spatial and temporal differentials, since the spatial solution is not generally continuous but instead is discretized.

We next focus on complex notation. Many vibration textbooks only use harmonic expressions (sine and cosine) for structural dynamic equations and solutions, but not only does complex notation make for a more succinct representation, it is actually required for fully generating solutions for multi-degree of freedom systems. The complex plane, also called the Argand diagram, is shown in Fig. 2.2.

**Supplementary Information** The online version contains supplementary material available at https://doi.org/10.1007/978-3-031-18207-5_2.

 A. M. Brown, *Structural Dynamics of Liquid Rocket Engines*, Synthesis Lectures on Mechanical Engineering, https://doi.org/10.1007/978-3-031-18207-5_2

**Fig. 2.1** Coordinate system
location and displacement
definitions

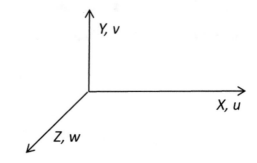

**Fig. 2.2** Argand diagram

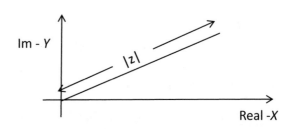

Where the complex value $z$, which we will generally identify with an overbar, is a
function of real x and imaginary y components.

$$\bar{z} = x + iy$$

$$|\bar{z}| = \sqrt{x^2 + y^2}$$

As harmonic analysis is an integral part of SD, we also can express these variables in
their polar form

$$x = r\cos\theta = |\bar{z}|\cos\theta$$
$$y = r\sin\theta = |\bar{z}|\sin\theta.$$

Finally, the Euler Identity enables the use of complex notation to express harmonic
characteristics

$$e^{i\theta} = \cos\theta + i\sin\theta$$

so,

$$\bar{z} = x + iy$$
$$= |\bar{z}|\cos\theta + i|\bar{z}|\sin\theta$$
$$= |\bar{z}|(\cos\theta + i\sin\theta)$$
$$\bar{z} = |\bar{z}|e^{i\theta}.$$

Other key mathematical concepts and procedures necessary for SD include basic linear algebra and Fourier analysis; it is left up to the reader to review those concepts as necessary. A small section covering Fourier analysis and orthogonality will be presented later.

Although a review of units may seem elementary to readers, it is a critical aspect of SD analysis, and it is frequently confused in the United States, where the Imperial system is still firmly entrenched in most industrial applications. Confusion in units caused the failure of the Mars Climate Orbiter Mission in 1999 when Lockheed-Martin Corporation, the prime contractor, sent NASA/JPL an engine impulse requirement in Imperial pound-second units without definition, and JPL expected the units to be SI newton-seconds. This caused an incorrect burn upon arrival at Mars, and the vehicle burned up in the atmosphere [1].

In the Imperial system, stiffness $k$ is defined as lb/in, where lb is a force. Mass $m$ is a derived unit, meaning we must derive a relationship between the values tabulated in material databases, which are in terms of force or weight $w$, to $m$. If we look at the simplest expression of Newton's second law, $F = m\,a$, where the acceleration on earth is one $g$ at sea level, equal to 386.09 in/sec$^2$, then

$$m = \frac{w\ lb}{g\ \frac{in}{sec^2}} = \frac{w\ lb - sec^2}{g\ \ \ in}.$$

As irksome as it may be, the best policy is to always state a mass term as "pound seconds squared per inch"; this term is informally referred to as a "slinch" or even as a "snail". At times mass is unfortunately referred to in units of lb-mass; this just means that the value is the mass of one pound under one g (or 1/386.1 lb seconds squared per inch), and so isn't constant. If the value of 1.0 is used in a natural frequency calculation when it should be 1/386.1, obviously the answer will be severely in error! The SI unit system is much easier in this regard as Kilograms mass is the basic unit and force in Newton's is the derived unit, so there is rarely any confusion.

Finally, as SD is heavily dependent on frequency analysis, it's worth emphasizing some basic frequency terminology and relationships. In mathematics, the "circular", or "angular" frequency $\omega$ is defined in radians/second, but in engineering applications, the "ordinary frequency" $f$ Hertz (cycles/second), or with rotating machinery, $N$ RPM (rotations per minute) are commonly used. To convert back and forth between $f$ and $\omega$ (make sure not to confuse Greek letter $\omega$ with Roman letter w), use

$$f\ Hz = \frac{\omega\ rad/sec}{2\pi} Hz$$

and then to get to RPM,

$$N RPM = (f\ Hz) * (60\,sec/min).$$

In addition, the conversion between frequency and period $T$ is

$$T = \frac{1}{f\,\mathrm{Hz}}\sec = \frac{2\pi}{\omega}\sec.$$

## 2.2  Free Vibration of Single Degree-of-Freedom Systems

In this section, we examine the free vibration of Single Degree-of-Freedom (SDOF) Systems. While this may seem academic, there are actually many systems (although not many structural systems) which can be accurately represented by a SDOF. The understanding of SDOF's is a basic building block for structural dynamics.

### 2.2.1  Modeling

The first step in any vibrations or structural dynamics text is mathematical modeling of the physical system. Consider an undamped spring-mass system and its free-body diagram shown in Fig. 2.3.

If an initial displacement $u$ is imposed onto the mass and released, the free-body diagram coupled with Newton's second law

$$\Sigma F_x = m\ddot{u}$$

yields the differential equation of motion

$$m\ddot{u} + ku = 0. \tag{2.1}$$

A nice example is a water bottle hanging from a series of rubber bands. This does introduce the gravity field, but if the bottle is slowly allowed to hang freely, reaching deflection $\delta$ in static equilibrium, before an initial displacement is introduced, the equation

**Fig. 2.3  a** Spring-mass system, **b** Free-body diagram

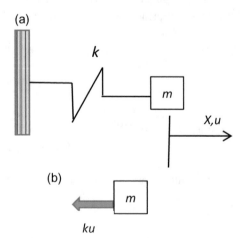

of motion shown above is completely valid. This can be easily proven using a free-body diagram, where the force of the mass in gravity $w$ is equal and opposite to the force from the static deflection $k\,\delta$. This property extends to almost all structural dynamics problems —a constant gravitational field can be ignored.

### 2.2.2  Solution of Equation of Motion

The solution to a differential equation consists of a transient and steady-state component. Although the transient portion does play a role in "dithered" excitation, in which the excitation frequency is not constant, the steady-state portion in SD problems in engines generally dominates. A very simple solution method will be presented therefore, which, although not entirely broadly applicable, yields the critical aspects needed.

We are interested in the solution for $u$ as a function of time $t$. From observation of SDOF spring-mass systems, we see that the response is harmonic, so we'll assume a solution of that form, and take the time derivative twice

$$
\begin{aligned}
u(t) &= A\cos(\omega t) \\
\dot{u}(t) &= -A\omega\sin(\omega t) \\
\ddot{u}(t) &= -A\omega^2\cos(\omega t)
\end{aligned}
\tag{2.2}
$$

where $A$ is some arbitrary amplitude. We can now substitute these expressions directly into the equation of motion, Eq. (2.1).

$$
m(-A\omega^2\cos\omega t) + k(A\cos\omega t) = 0
$$
$$
A\cos\omega t(k - \omega^2 m) = 0
$$

In order for this equation to be true, either

$$
A\cos\omega t = 0
$$

or

$$
k - \omega^2 m = 0.
$$

But since $\cos(\omega t)$ is only 0 for specific values of $t$, for the first equation to be true, $A$ would have to equal zero. This would yield no response (and we know there is a response from observation), and therefore is the "trivial" solution. This means the second part must be true, so solving for $\omega^2$, which is defined to be the eigenvalue $\lambda$,

$$
\omega^2 = \frac{k}{m} \equiv \text{eigenvalue } \lambda.
$$

and

$$\omega = \sqrt{\frac{k}{m}} \, \text{Rad/sec} \tag{2.3}$$

Where $\omega$ is called the natural frequency. Equation (2.3) is probably the most critical single formula in structural dynamics, well worth memorizing. But it is not actually the solution we are looking for, which is what is $u$ doing in time. That solution is obtained by plugging the natural frequency into our original harmonic assumption in Eq. (2.2), resulting in

$$u(t) = A \cos\left(\sqrt{\frac{k}{m}}t\right).$$

What is critical yet difficult to intuitively understand is that the amplitude of vibration has an arbitrary amplitude $A$. Free vibration means there are no external stimuli, so the motion only occurs when some kind of initial condition (displacement, velocity, acceleration, or even an initial stress state) is applied to the system and then released. Only then will the system vibrate at its natural frequency. For example, with an initial displacement of

$$u(t = 0) = 1.5$$

then

$$A \cos\left[\sqrt{\frac{k}{m}}(0)\right] = 1.5$$

$$A * (1) = 1.5$$

$$A = 1.5,$$

so the steady state response of an undamped, SDOF system with an initial displacement of 1.5 is

$$u(t) = 1.5 \cos\sqrt{\frac{k}{m}}t.$$

If the initial displacement is defined as $u_o$, and an initial velocity $\dot{u}_o$ is also applied, the assumed solution must be expressed as a sum of a sine and cosine terms (or the complex form can be used), and after some simple algebra, the solution of $u(t)$ is

$$u(t) = u_o \cos(\omega t) + \frac{\dot{u}_o}{\omega} \sin \omega t \tag{2.4}$$

*Example 2.1* A 386.1 lb payload in the shuttle cargo bay was unexpectedly jostled during a test and just barely tapped up against an adjacent panel resting 3″ away as it vibrated.

**Fig. 2.4** Example 2.1
Displacement versus time

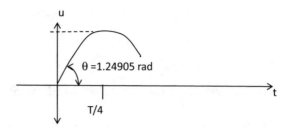

There was a strain gage on the payload which failed, but its initial reading could be transformed to yield an initial angle of the displacement vs. time curve of θ = 1.24905 Radians. Always ready to seize the opportunity to get useful data even from an incident, you realize that if you assume no damping, you can use the undamped equation for free vibration of a SDOF system to obtain the stiffness $k$ of the payload hold-down trunnions, which could come in handy in a host of ways.

**Solution** The displacement versus time plot is shown in Fig. 2.4.

Applying Eq. (2.4) for displacement with an initial velocity, and applying our known initial conditions,

$$u(t) = u_o \cos(\omega t) + \frac{\dot{u}_o}{\omega} \sin \omega t$$

$$u_o = 0; \; \dot{u}_o = \frac{du}{dt} = \tan \theta = \tan(1.29405) = 3$$

Therefore, $u(t) = \frac{3}{\omega} \sin \omega t$.

Now, we also know the maximum deflection during the first cycle of vibration is 3 in. For sin waves, this maximum occurs at a time equal to one quarter of a period, so

$$u\left(\frac{T}{4}\right) = \sqrt{3} = \frac{3}{\omega} \sin \omega \frac{T}{4}$$

$$\sqrt{3} = \frac{3}{\omega} \sin\left(\frac{\pi}{2}\right)$$

Therefore,

$$\sqrt{3} = \frac{3}{\omega} \rightarrow \omega = \sqrt{3}$$

$$m = \frac{w}{g} = \frac{386.1 lb}{386.1 \frac{in}{s^2}} = 1 \frac{lb - s^2}{in}$$

and applying the formula for natural frequency $\omega = \sqrt{\frac{k}{m}}$, we can obtain the value of stiffness,

$$\sqrt{3} = \sqrt{\frac{k}{1}} \rightarrow k = 3\frac{lb}{in}.$$

### 2.2.3   Damped Free Vibration of SDOF Systems

Any realistic system, of course, has some level of energy loss, which is generally lumped in structural dynamic analysis into a linear damping term proportional to velocity. Damping in SD is generally neither linear nor proportional to velocity, but it is close enough for this assumption to be valuable, especially for design purposes. The inaccuracy of this assumption should be kept in mind, though, for any kind of failure analysis or uncertainty assessment. Figure 2.5 shows a damped SDOF system, free-body diagram, and resulting equation of motion.

A convenient technique for solving this equation of motion is to apply the Laplace transform, which is quite similar to the Euler transform discussed in the mathematics section. Using this technique, we assume the solution has the form:

$$u(t) = Ue^{st}$$
$$\dot{u}(t) = sUe^{st}$$
$$\ddot{u}(t) = s^2Ue^{st}$$

If these are substituted into the equation of motion,

$$m\left(s^2Ue^{st}\right) + c(sUe^{st}) + k(Ue^{st}) = 0$$
$$Ue^{st}\left(ms^2 + cs + k\right) = 0$$

So the non-trivial solution is

$$ms^2 + cs + k = 0$$

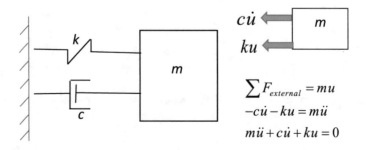

**Fig. 2.5**  Sketch and free-body diagram of SDOF system with damping

which is a simple quadratic, so the solution for $s$ is

$$s_{1,2} = \frac{-c \pm \sqrt{c^2 - 4mk}}{2m}$$

$$= \frac{-c}{2m} \pm \sqrt{\left(\frac{c}{2m}\right)^2 - \frac{k}{m}}.$$

This solution gives three categories of damping. The first is critical damping, where

$$\left(\frac{c}{2m}\right)^2 - \frac{k}{m} = 0$$

$$\frac{c^2}{4m^2} = \frac{k}{m}$$

$$c^2 = \frac{4m^2 k}{m}$$

$$c = 2\sqrt{km} \equiv c_{critical}$$

The second category is overdamped, where

$$\left(\frac{c}{2m}\right)^2 - \frac{k}{m} > 0$$

$$c > 2\sqrt{km}$$

$$c > c_{critical}$$

The third category, which is the case virtually all the time in SD systems, is underdamped.

$$\left(\frac{c}{2m}\right)^2 - \frac{k}{m} < 0$$

$$c < 2\sqrt{km}$$

$$c < c_{critical}.$$

Figure 2.6 shows the oscillatory response obtained by substituting values for c into the assumed solution that fit into these three damping categories.

We now will go through the steps to derive the temporal solution for an underdamped system to highlight several important relationships and equations as well as the final response form. From the damped equation of motion

$$m\ddot{u} + c\dot{u} + ku = 0,$$

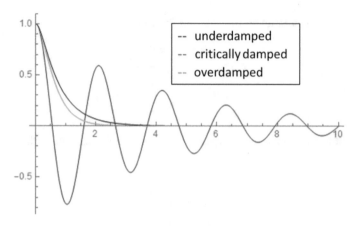

**Fig. 2.6** Damped SDOF Response for different levels of damping

divide by m

$$\ddot{u} + \frac{c}{m}\dot{u} + \frac{k}{m}u = 0.$$

Now define the viscous damping ratio $\zeta$ as the percentage of critical damping for the system,

$$\zeta = \frac{c}{c_{\text{critical}}}.$$

As an aside, $\varsigma \approx 0.5\eta$, where $\eta$ is defined to be the "structural damping loss factor", which is frequently used instead of $\zeta$ in SD applications. Using $\omega^2 = \frac{k}{m}$, we get

$$\ddot{u} + \frac{\zeta c_{cr}}{m}\dot{u} + \omega^2 u = 0$$

Then use $c_{cr} = 2\sqrt{km}$ to get

$$\ddot{u} + \frac{2\zeta\sqrt{km}}{m}\dot{u} + \omega^2 u = 0$$
$$\ddot{u} + 2\zeta\omega\dot{u} + \omega^2 u = 0$$

(2.5)

The form of Eq. (2.5) is very important; all the parameters (coefficients of the time-varying terms) are basic dynamic characteristics of the structure, meaning that a SDOF system can be characterized entirely by these somewhat easy-to-measure properties. This same property will be extended to multi-degree-of-freedom structures later and is critical for MDOF structural dynamic analysis.

Continuing the derivation, we assume a harmonic solution in the Lagrange complex form

$$u = U e^{st}$$

And plug this into Eq. (2.5) to give

$$U e^{st} \left( s^2 + 2\zeta\omega s + \omega^2 \right) = 0$$

And, like before,

$$s_{1,2} = \frac{-2\zeta\omega \pm \sqrt{(2\zeta\omega)^2 - 4\omega^2}}{2}$$

$$= -\zeta\omega \pm \omega\sqrt{\zeta^2 - 1}$$

$$= -\zeta\omega \pm i\omega\sqrt{1 - \zeta^2}$$

The coefficient of the imaginary term is defined to be the damped natural frequency, $\omega_d$; free vibration will be at this frequency. We now plug each solution back into the assumed solution

$$u = U_1 e^{s_1 t} + U_2 e^{s_2 t}$$

$$= U_1 e^{\left(-\zeta\omega + i\omega\sqrt{\zeta^2 - 1}\right)t} + U_2 e^{\left(-\zeta\omega - i\omega\sqrt{\zeta^2 - 1}\right)t}$$

$$= e^{-\zeta\omega t} \left( U_1 e^{i\omega_d t} + U_2 e^{-i\omega_d t} \right)$$

And, similar to before, using the Euler Identity

$$= e^{-\zeta\omega t} \left( A\cos\omega_d t + A\sin\omega_d t \right)$$

Now apply the initial conditions, first for displacement

$$u(t = 0) = u_o$$

$$e^0 (A\cos 0 + B\sin 0) = u_o$$

$$A = u_o$$

Then for velocity

$$\dot{u}(t = 0) = \dot{u}_o$$

and differentiate with respect to $t$:

$$e^{-\zeta\omega t}(-\omega_d A \sin \omega_d t + \omega_d B \cos \omega_d t)$$
$$- \zeta\omega e^{-\zeta\omega t}(A \cos \omega_d t + B \sin \omega_d t) = \dot{u}(t)$$

Substitute $t = 0$:

$$\omega_d B - \zeta\omega A = \dot{u}_o$$
$$B = \frac{\dot{u}_o + \zeta\omega u_o}{\omega_d}$$

We therefore arrive at the final solution for the free vibration of damped SDOF system.

$$u(t) = e^{-\zeta\omega t}\left(u_o \cos \omega_d t + \frac{\dot{u}_o + \zeta\omega u_o}{\omega_d} \sin \omega_d t\right) \tag{2.6}$$

Figure 2.7 is a graph of the time history; note the graphical manifestations of the two initial conditions and the shape of the exponential envelope of the response.

The damped natural period is

$$T_d = \frac{2\pi}{\omega_d \text{Rad/sec}} = \frac{1}{f \text{ Hz}}$$

and will always be longer than $T$ (the undamped natural period). It should be noted that in SD applications, $\zeta$ is almost always less than 10%, and in rocket engine applications is less than 0.5%! if we take a closer look at the formulation for the $\omega_d$ for these values,

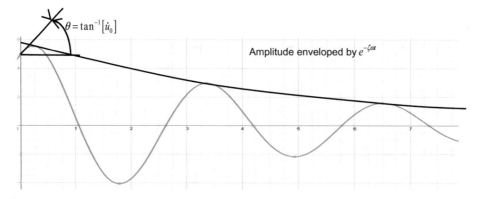

**Fig. 2.7** Damped SDOF temporal response, showing IC's and amplitude envelope

$$\omega_d = \omega\sqrt{1 - 0.1^2} = 0.99\omega$$
$$\omega_d = \omega\sqrt{1 - 0.01^2} = 0.998\omega$$

it becomes clear that there is only a negligible difference between $\omega_d$ and $\omega$, which at least takes away one headache!

### 2.2.4 Measurement of Damping Using Log Decrement Method

This is one of the most common methods for measuring damping in a SD system. An initial displacement is imposed on the system and released, and the response $u_1(t_1)$ and $u_2(t_2)$ is measured at successive peaks, as shown in Fig. 2.6. Since the peaks lie very close to the exponential envelope, we can just use that part of the solution

$$\frac{u_1(t_1)}{u_2(t_2)} = \frac{e^{-\zeta\omega t_1}}{e^{-\zeta\omega t_2}} = e^{-\zeta\omega(t_1 - t_2)}$$
$$= e^{\zeta\omega(t_2 - t_1)}$$
$$= e^{\zeta\omega T_d}$$

where

$$T_d = \frac{1}{f_d} = \frac{2\pi}{\omega_d} = \frac{2\pi}{\omega\sqrt{1 - \zeta^2}}.$$

so

$$\frac{u_1}{u_2} = \exp\left(\zeta\omega\frac{2\pi}{\omega\sqrt{1 - \zeta^2}}\right).$$

Therefore, define the Log Decrement $\delta$ as

$$\delta \equiv \ln\left(\frac{u_1}{u_2}\right) = \frac{\zeta 2\pi}{\sqrt{1 - \zeta^2}}$$

and if $\zeta \gg 1$, $\sqrt{1 - \zeta^2} \cong 1$, then

$$\ln\left(\frac{u_1}{u_2}\right) = \zeta 2\pi, \quad \rightarrow \quad \zeta = \frac{\ln(u_1/u_2)}{2\pi} = \frac{\delta}{2\pi}. \tag{2.7}$$

In many turbine and other aeromechanical applications, the metric of interest is actually $\delta$ instead of $\zeta$.

Rao has collated a useful table of typical values of $\zeta$ for engineering materials as well as built-up structures [2]. Applicable values for LRE's are $< 0.01\%$ for material damping, $0.1\%$ for turbine blades in fir trees (to be discussed in Chap. 3), $0.5\%$ for turbine blades with dampers, 2–8% for impeller blades in liquid oxygen, 1–4% for welded construction, and 3–10% for bolted construction.

## 2.3   Forced Response of SDOF Systems

### 2.3.1   Harmonic Excitation

The solution for the response of SDOF systems to harmonic excitation is a critical building block for most SD calculations. However, in rocket engine turbopumps, this harmonic response calculation is critical in of itself, as there are huge sources of harmonic excitation which have caused numerous failures.

Returning to our SDOF system, we now apply a force $P = F_o\cos(\Omega t)$, as shown in Fig. 2.8. We use capital omega $\Omega$ as the excitation frequency, and it is critical to understand it is not the same as the natural frequency little omega $\omega$. The applied force $P$ is a pure force, like an aerodynamic force, not an enforced displacement as is generally the more intuitive case. The equation of motion is now

$$m\ddot{u} + c\dot{u} + ku = F_o \cos \Omega t \tag{2.8}$$

The solution to this second order ordinary differential equation will have two parts that are very important in forced response, the particular (more descriptively, the nonhomogeneous, steady state) solution, and the complimentary (the homogeneous, transient) solution. Nonhomogeneous refers to the component of the solution resulting from a non-zero right-hand-side of the equation of motion, i.e., an external force, and the response continues for as long as the excitation is applied, hence it is steady-state.

Homogeneous means the right-hand side is zero, and the only response is due to an initial condition which dies away with any non-zero amount of damping, so it is therefore transient, or temporary. This part of the solution has already been derived as Eq. (2.4). However, any change in steady-state harmonic portion of the excitation affects the system in the same way as an initial condition. Therefore, an excitation that is defined in the temporal domain (as opposed to the frequency domain) is termed in engineering parlance as "transient", referring to the fact that a portion of the response will fade away (assuming

**Fig. 2.8** Harmonic Excitation of SDOF

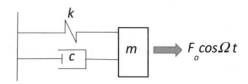

the time history is not just a simple harmonic). Since most of the response will not fade away (the portion that is harmonic), this term is somewhat misleading. SD analysis of rocket engines is far is more concerned with the steady-state portion of the response, as that causes the destructive effect of resonance.

Focusing now on the steady state solution, we will now apply the Euler Identity and start working in the complex plane. As before, a solution is of the form $\bar{u}(t)_{ss} = \bar{U}e^{i\Omega t}$ where $\bar{u}(t)_{ss}$ is a complex number containing phase (relationship of the response amplitude to the force amplitude). Differentiating as before,

$$\dot{u}(t) = \bar{U}i\Omega e^{i\Omega t}$$
$$\ddot{u}(t) = -\bar{U}\Omega^2 e^{i\Omega t}$$

and let excitation be represented (for convenience) by

$$f(t) = F_o(\cos \Omega t + i \sin \Omega t) = F_o e^{i\Omega t}$$

where the imaginary part will be 0.

Now, substituting into the equation of motion,

$$m\bar{U}\Omega^2 e^{i\Omega t} + c\Omega \bar{U}i e^{i\Omega t} + k\bar{U}e^{i\Omega t} = F_o e^{i\Omega t}$$
$$\underbrace{\left(-m\Omega^2 + c\Omega i + k\right)}_{Z(i\Omega) = \text{MechanicalImpedence}} \bar{U} = F_o$$

$$\therefore \quad \bar{U} = \left(F_o / - m\Omega^2 + k + ic\Omega\right) \cdot \frac{\frac{1}{k}}{\frac{1}{k}}$$
$$= \left(\frac{F_o}{k} / - \frac{m}{k}\Omega^2 + 1 + i\frac{c}{k}\Omega\right).$$

Now, using the equivalences derived to obtain Eq. (2.5), repeated here,

$$\omega^2 = \frac{k}{m} \rightarrow \frac{m}{k} = \frac{1}{\omega^2}$$
$$c/m = 2\zeta\omega \rightarrow c = m2\zeta\omega \rightarrow$$
$$c/k = \frac{m}{k}2\zeta\omega = \frac{2\zeta\omega}{\omega^2} = \frac{2\zeta}{\omega}$$

we obtain

$$\bar{U} = \frac{Fo/k}{\left(1 - \left(\frac{\Omega}{\omega}\right)^2\right) + i2\zeta\frac{\Omega}{\omega}}.$$

Define the frequency ratio $r = \frac{\Omega}{\omega}$, then

$$\overline{U} = \frac{F_o/k}{(1 - r^2) + i2\zeta r}.$$

Now, define the static response $U_{st}$ resulting from the application of force $F_o$ using

$$kU_{st} = F_o \quad \rightarrow \quad U_{st} = F_o/k,$$

then we can define the "Complex Frequency Response"

$$\overline{H}(\Omega) = \frac{\text{Dynamic Response } \overline{U}}{\text{Static Response } \overline{U}_{st}}$$

so substituting the expressions for $U_{st}$ and $\overline{U}$, we get

$$\overline{H}(\Omega) = \frac{1}{(1 - r^2) + i2\zeta r} \quad \rightarrow \quad \overline{U} = \overline{H}(\Omega)\overline{U}_{st}. \tag{2.9}$$

A great deal of useful information can be obtained by examining the Real and Imaginary parts of $\overline{H}(\Omega)$ in depth. To enable this, multiply the numerator and denominator by the complex conjugate of the denominator

$$\overline{H}(\Omega) = \frac{1}{(1 - r^2) + i2\zeta r} \cdot \frac{(1 - r^2) - i2\zeta r}{(1 - r^2) - i2\zeta r}$$

$$\overline{H}(\Omega) = \underbrace{\frac{(1 - r^2)}{(1 - r^2)^2 + (2\zeta r)^2} + i\frac{-2\zeta r}{(1 - r^2)^2 + (2\zeta r)^2}}_{\text{of form } x + iy = Ae^{i\phi}}$$

$$\text{where } A = \sqrt{x^2 + y^2} \text{ and } \tan \phi = \frac{y}{x}.$$

Remember, the final solution we are looking for is in the form

$$\overline{u}_{ss}(t) = \overline{U}e^{i\Omega t}$$

which can be expressed as

$$\overline{u}_{ss}(t) = \overline{U}_{st}\overline{H}(\Omega)e^{i\Omega t}$$
$$= \overline{U}_{st}|\overline{H}(\Omega)|e^{i\phi}e^{i\Omega t} = \overline{U}_{st}|\overline{H}(\Omega)|e^{i(\Omega t + \phi)} \tag{2.10}$$

Where the magnitude of the Complex Frequency Response $H$ is

$$|\overline{H}(\Omega)| = \sqrt{\frac{(1-r^2)^2 + (2\zeta r)^2}{\left((1-r^2)^2 + (2\zeta r)^2\right)^2}}$$

$$= \sqrt{\frac{1}{(1-r^2)^2 + (2\zeta r)^2}}$$

and where the phase lag $\phi = \tan^{-1}\left(\frac{-2\zeta r}{1-r^2}\right)$. The magnitude of the complex frequency response is also called the magnification M, and we can simply say that the magnitude of the dynamic response is M times the magnitude of the static response.

$$|\overline{U}| = M|\overline{U}_{static}|$$

We now can define one of the most critical concepts in structural dynamics; resonance is the condition at which $\Omega = \omega$, i.e., r = 1.

$$|\overline{H}(\Omega)| = \frac{1}{2\zeta}. \tag{2.11}$$

The value of the magnitude of $H$ at r = 1 is also called the "Quality Factor", which is very commonly used in all areas of SD.

The phase lag also merits some discussion. This is the wave angle relationship between the excitation wave and the response wave. A plot of the phase lag versus $r$ is shown in Fig. 2.9. It's worth examining how the lag approaches a value of exactly 90°, or ¼ of a wave, at resonance, and then above resonance approaches 180°, which is completely out-of-phase with the excitation.

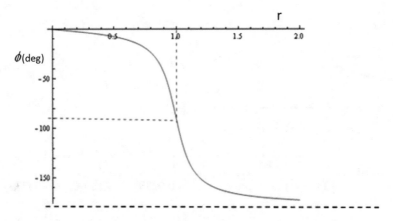

**Fig. 2.9** Phase angle versus frequency ratio

It should also be noted that many software programs require the use of the x, y form of ArcTan to account for the correct quadrant (whether the numerator and denominator are individually positive or negative). This form will be

$$\phi = \text{ArcTan}(1 - r^2, -2\zeta r).$$

Since SD is concerned in general with functions of time, it's also worth expressing the phase lag in terms of time. The time lag for a given phase lag can be derived as

$$t_{lag} = \frac{\phi}{2\pi}T = \frac{\phi}{2\pi}\frac{1}{f} = \frac{\phi}{2\pi}\frac{2\pi}{\Omega} = \frac{\phi}{\Omega} \text{ sec.}$$

*Example  F = 2; c = 0.6; m = 1; k = 9.*

$$\omega = \sqrt{\frac{k}{m}} = 3$$

$$\zeta = \frac{c}{2\sqrt{km}} = 0.1$$

$$U_{static} = \frac{F_o}{k} = 0.222$$

At resonance, $|U| = Q\,U_{static}$

$$\therefore |U| = \frac{1}{2\zeta}(0.2222) = 1.111$$

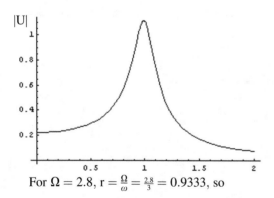

For $\Omega = 2.8$, $r = \frac{\Omega}{\omega} = \frac{2.8}{3} = 0.9333$, so

$$|\overline{H}(\Omega)| = \sqrt{\frac{1}{(1 - r^2)^2 + (2\zeta r)^2}} = \sqrt{\frac{1}{(1 - 0.93333^2)^2 + (2*0.1*0.93333)^2}}$$

$$= 4.408$$

$$\phi = \tan^{-1}\left[\frac{-2\zeta r}{1-r^2}\right] = \tan^{-1}\left[\frac{-2(0.1)0.933}{1-0.933^2}\right]$$

$$= \tan^{-1}\left(-\frac{0.0187}{0.129}\right) = -0.9665$$

### 2.3.2   Examination of Real and Imaginary Parts of the Complex Frequency Response

Using the Euler identity, $e^{i\theta} = \cos\theta + i\sin\theta$, and combining the terms in the exponents, Eq. (2.11) becomes:

$$\bar{u}_{ss}(t) = \frac{F_o}{k}\left|\overline{H}(\Omega)\right|\cos(\Omega t + \phi) + i\frac{F_o}{k}\left|\overline{H}(\Omega)\right|\sin(\Omega t + \phi).$$

The actual displacement as a function of time is equal to just the real part of the complex function $\bar{u}_{ss}(t)$, but we need the imaginary part as well to know the full characteristics of the response in the frequency domain.

The magnitude and real and imaginary parts are shown in Fig. 2.10. The imaginary part is quite useful as its peak occurs at a frequency equal to the exact value of resonance, unlike both the magnitude and the real part.

### 2.3.3   Measurement of Damping Using Half-Power Method

At this point we have enough information to discuss the most common method for obtaining damping from test data, the "Half-Power" method. If a frequency response curve of a system is obtained, the first step is to determine the undamped natural frequency $f_i$ and the resonant response amplitude $|U|$. Using terminology from electrical

**Fig. 2.10** Real and imaginary parts of magnification

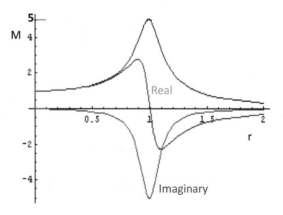

systems, we then obtain the frequencies which are at amplitudes equal to $\sqrt{2}/2$, or 70.7% of the magnitude. Applying the derivation in the text by Craig previously referenced,

$$\zeta_i = \frac{1}{2}\frac{f_2 - f_1}{f_i}. \tag{2.12}$$

For example, given a frequency response as shown in Fig. 2.11, the associated calculations can be made.

## 2.3.4  Support Motion

The response calculations presented so far have been for a pure force applied to a mass. In practice, the source of the excitation frequently is motion of the base of the system, as shown in Fig. 2.12. The two quantities generally of interest are the relative displacement of the mass from the base, which will correspond to the strain or stress, and the acceleration of the mass.

The displacement of the base $d$ is assumed to be harmonic of the form

$$d = De^{i\Omega t}$$
$$\ddot{d} = -D\Omega^2 e^{i\Omega t}$$

The relative displacement is $s = u - d$, and the total displacement $u = d + s$.

A free-body diagram shown in Fig. 2.13 is critical for understanding the forces on $m$.

If we're interested in the relative displacement, our equation of motion can be derived as

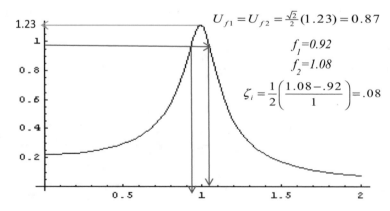

$$U_{f1} = U_{f2} = \frac{\sqrt{2}}{2}(1.23) = 0.87$$
$$f_1 = 0.92$$
$$f_2 = 1.08$$
$$\zeta_i = \frac{1}{2}\left(\frac{1.08 - .92}{1}\right) = .08$$

**Fig. 2.11**  SDOF frequency response for half power calculation

**Fig. 2.12** Base excitation of SDOF system

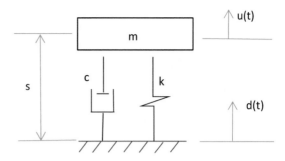

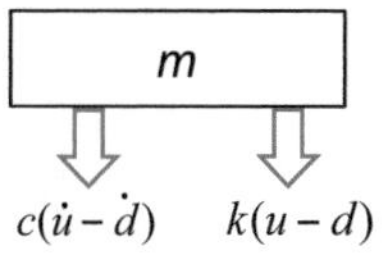

**Fig. 2.13** Free body diagram for relative displacement

$$m(\ddot{u}) + c(\dot{u} - \dot{d}) + k(u - d) = 0$$
$$m(\ddot{d} + \ddot{s}) + c\dot{s} + ks = 0$$
$$m\ddot{s} + c\dot{s} + ks = -m\ddot{d} = mD\Omega^2 e^{i\Omega t}$$

so following the same procedure as in the last section, that is, for an assumed solution of s in the form $s = \overline{S}e^{i\Omega t}$, we can derive the "relative motion frequency response function",

$$\overline{S}e^{i\Omega t}\left(-m\Omega^2 + ic\Omega + k\right) = m\overline{D}\Omega^2 e^{i\Omega t}$$
$$\frac{\overline{S}}{\overline{D}} = \frac{m\Omega^2}{(k - m\Omega^2) + ic\Omega} \Rightarrow$$
$$\frac{\overline{S}}{\overline{D}} = \frac{r^2}{(1 - r^2) + i2\zeta r}$$
$$\left|\frac{\overline{S}}{\overline{D}}\right| = r^2\left|\overline{H}(\Omega)\right|.$$

The subtle but important difference from the Frequency Response function for the force excitation system is seen in Fig. 2.14.

The equation for the phase angle, however, is the same as for a force excitation

$$\varphi = \tan^{-1}\left(\frac{-2\zeta r}{1 - r^2}\right)$$

Valuable insights on the general behavior are that for $r$ much less than 1, i.e., excitation frequency much less than the natural frequency, the relative response is very small. At $r = 1$, we have a resonant response, where the peak is $1/2\zeta$, just like the force frequency

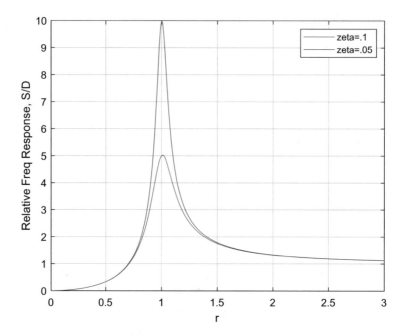

**Fig. 2.14** Relative complex frequency response

response, but interestingly, at high $r$, the response approaches 1, but is completely out of phase. The result of this is that the base will be moving, but the object remains still in space. As mentioned previously, this can be seen clearly with the water bottle on rubber band experiment.

Occasionally we are interested in the absolute motion, e.g., an astronaut in a capsule on top of a launch vehicle. We start with the same equation of motion

$$m\ddot{u} + c\dot{u} + ku = c\dot{d} + kd$$

and using the same technique, derive

$$\frac{|U|}{|D|} = \sqrt{1 + (2\zeta r)^2} \, \frac{1}{\sqrt{(1 - r^2)^2 + 2\zeta r^2}}$$

$$= \sqrt{1 + (2\zeta r)^2} |H(\Omega)|$$

which also happens to equal the relationship between the force transmitted to a static base from a vibrating system divided by a statically transmitted force, a quantity called the transmissibility, which is of interest in designing the mounts of rotating machinery such as turbopumps.

$$\frac{F_{transmitted\ dynamically}}{F_{statically}} \equiv \text{Transmissibility}$$

### 2.3.5  SDOF Response to Arbitrary and Periodic Excitation

Applied forces are not, in general, perfect harmonic (sine and cosine) waves. We can deal with this fact in two ways, either representing the forces as a sum of harmonics and then sum the response of each one, or calculate the response to the actual temporal (time history) loading using the "impulse response function." Although the second technique is academically elegant, it is rarely used explicitly, although it is "under the hood" in finite element codes, so an explanation will be left to more comprehensive SD textbooks. An understanding of the first method, called the spectral, or frequency domain approach, is critical for all levels of SD analysis of LRE's.

Assume that there is a periodic but non-harmonic excitation as shown in Fig. 2.15. where

$p(\tau) = p(\tau + T).$

Jean Fourier (1768–1830) discovered that $p$ can be written as a sum of the average, cosines, and sines.

$$p(t) = a_o + \sum_{n=1}^{\infty} [a_n \cos(n\Omega_1 t) + b_n \sin(n\Omega_1 t)]$$

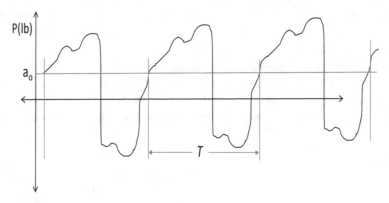

**Fig. 2.15** Periodic but non-harmonic signal

where

$$a_0 = \frac{1}{T_1} \int\limits_{\tau}^{\tau+T} p(t)dt = \text{avg value of p(t)}$$

$$a_n = \frac{2}{T_1} \int p(t) \cos(n\Omega_1 t)dt$$

$$b_n = \frac{2}{T_1} \int p(t) \sin(n\Omega_1 t)dt$$

For example, using the Fourier Series, a square wave, shown in Fig. 2.16, can be expressed as

$$p(t) = \frac{4p_0}{\pi} \sum_{n=1,3,5,\dots} \left(\frac{1}{n}\right) \sin(n\Omega_1 t)$$

Turbopump excitations all take this form, so we will revisit this in detail in Chap. 3.

### 2.3.6 Shock Response Spectra

Our final area of focus within forced response of SDOF systems is called the Shock Response Spectra (SRS); this methodology is used in both rocket engines and throughout launch vehicles to help define requirements for components that will have to withstand high vibration environments and short time scales. The given excitation is usually mechanically transmitted through the base. Although this methodology is mostly used for spacecraft and launch vehicle component assessments, the concepts are important for engine system loads analysis as well. It particularly is valuable if many components have to withstand the same base excitation environment.

The SRS gives a curve consisting of peak acceleration responses of sequential SDOF systems which have natural frequencies equal to every component of the excitation frequency range for a given value of Q, the Quality Factor $1/2\zeta$, for a specific acceleration base input (see Fig. 2.17). In essence therefore, it's a measure of how a perfectly resonant

**Fig. 2.16** Square wave periodic function

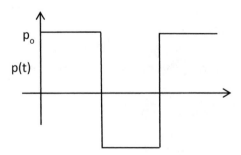

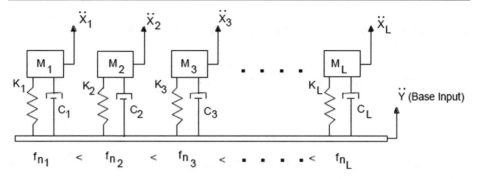

**Fig. 2.17**  Schematic of shock response spectra system

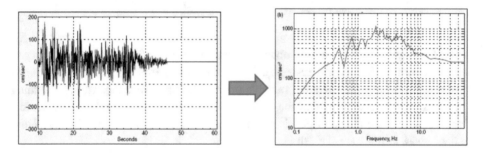

**Fig. 2.18**  Acceleration excitation and corresponding SRS

SDOF would respond to a given excitation, and so is a measure of the worst-case response to the excitation. A component can be "qualified" for a particular excitation environment by shaking it at levels equal to the SRS plus some agreed upon margin; if it survives, then it will definitely be ok in operation.

For example, a ground acceleration trace of a 1940 earthquake in El Centro, Ca, is shown along with its corresponding SRS in Fig. 2.18 [3].

## 2.4  Modal Analysis of Multiple DOF Systems

As simplified as it may seem, the single degree of freedom theory discussed up to this point has a great many practical applications, in particular describing mechanical systems, and including extremely interesting phenomena like machine tool chatter. However, since we focus in this text on predicting the response of rocket engine structures, the SDOF method is inadequate for this purpose. Although generally impractical for application, we will briefly examine a continuous representation of these structures, in which the mass and stiffness are spread smoothly throughout the structure. This will help us gain an understanding of basic structural dynamic characteristics and how they are connected to the eigenvalue problem studied in Calculus.

### 2.4.1   Continuous System Equation of Motion and Solution

Let's examine the equation of motion for a continuous string, which is a second order partial differential equation in time and space also known as the wave equation,

$$\left(\frac{P}{\rho A}\right)\frac{\partial^2 u}{\partial x^2} = \frac{\partial^2 u}{\partial t^2}$$

where $P$ is the forcing function, $\rho$ is density, $A$ is cross-sectional area, $u$ is displacement, $x$ is location, and $t$ is time.

The solution has two parts, the temporal (time) solution, and the spatial (space) solution

$$u_i(x,t) = A_i \sin\left(\frac{i\pi x}{l}\right)\cos\left(\frac{i\pi P}{\rho A l}t + \alpha\right)$$

where $l$ = length, $A_i$ is a constant obtained from initial conditions, and $\alpha$ is the phase angle between excitation and response. This solution is extremely useful for stringed instruments, of course, but here it's valuable as a basis of understanding. The critical thing to note is that the spatial part of the solution is the natural mode, or the mathematical eigenvector, and the temporal part is characterized by the natural frequencies, or eigenvalues, where in this case the $i$'th eigenvalue $\lambda_i$ is

$$\lambda_i = \frac{i\pi P}{\rho A l}$$

Excellent visualizations of the mode shapes vibrating at their respective natural frequencies are on a web site published by Penn State, www.acs.psu.edu/drussell/demos/string/fixed.html. There are analytical expressions for the eigenvalues and eigenvectors of simple continuous structures obtained by the solution of these types of partial differential equations, and these are valuable for "sanity checks", first cut quick analysis, and general understanding, but rarely are rocket engine structures simple enough to be adequately characterized using the continuous solutions. An excellent collection of these solutions is presented in the handbook by Blevins [4].

### 2.4.2   Generating Equations of Motion for Multiple Discrete DOF Systems

The structural dynamic analysis of rocket engine systems requires the use of the finite element method, which is a technique for solving the equations describing the response of multiple discrete degree-of-freedom systems. In this technique, the true continuous structure is broken down mathematically into hundreds of thousands of nodes, which each have between 3 and 6 degrees of freedom.

For the purpose of gaining a basic theoretical understanding of the dynamics of MDOF systems, we will use again use Newton's method for generating the equations of motion, as we did for SDOF systems. This method works well for translational systems but falls apart if rotational degrees of freedom are included, in which case Lagrangian energy methods must be used. However, as the equations are generally derived "under the hood" of software codes, we can focus on the simpler method to provide the necessary basic understanding.

The procedure for determining the basic structural dynamic characteristics (natural frequencies, mode shapes) is not trivial, and it best to follow a step-by-step procedure. Textbooks usually skip many of these following steps as being "intuitively obvious", which they are not. We will present several different examples illustrating different aspects of this process, as it is central to the discipline of SD. Part A of that procedure, which is to obtain the matrix equations of motion, is as follows:

1. Draw Free Body Diagram (FBD) for each mass
2. For each DOF, write sum of External Forces = mass * acceleration, in this order

$$m\ddot{u}_n = \sum F_n$$

3. Very carefully, expand above equation completely, putting only external forces on right hand side (RHS)
4. Group by degree of freedom, making sure coefficient has the correct sign.
5. Write the equation of motion in matrix format—should be symmetric.

As this is such a critical technique for understanding modal analysis, the entire procedure for the 3-DOF system in Fig. 2.19 is shown here.

Degree of Freedom 1:

1. $k_1 u_1 \longleftarrow$ [m1] $\longrightarrow k_2(u_2 - u_1)$ with $u$ shown above
2. $m_1\ddot{u}_1 = -k_1 u_1 + k_2(u_2 - u_1)$
3. $m_1\ddot{u}_1 + k_1 u_1 - k_2 u_2 + k_2 u_1 = 0$
4. $m_1\ddot{u}_1 + (k_1 + k_2)u_1 + (-k_2)u_2 = 0$

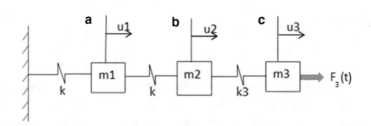

Fig. 2.19 a 3-DOF system; b Free body diagram DOF 1; c Free body diagram DOF2

Degree of Freedom 2:

1.

2. $m_2\ddot{u}_2 = -k_2(u_2 - u_1) + k_3(u_3 - u_2)$
3. $m_2\ddot{u}_2 + k_2u_2 - k_2u_1 - k_3u_3 + k_3u_2 = 0$
4. $m_2\ddot{u}_2 + (-k_2)u_1 + (k_2 + k_3)u_2 + (-k_3)u_3 = 0$

Degree of Freedom 3:

2. $m_3\ddot{u}_3 = -k_3(u_3 - u_2) + F_3$
3. $m_3\ddot{u}_3 + k_3u_3 - k_3u_2 = F_3$
4. $m_3\ddot{u}_3 + (-k_3)u_2 + k_3u_3 = F_3$

Step 5. The last equation in each set goes directly into the Matrix Equation,

$$\begin{bmatrix} m_1 & 0 & 0 \\ 0 & m_2 & 0 \\ 0 & 0 & m_3 \end{bmatrix} \begin{Bmatrix} \ddot{u}_1 \\ \ddot{u}_2 \\ \ddot{u}_3 \end{Bmatrix} + \begin{bmatrix} k_1 + k_2 & -k_2 & 0 \\ -k_2 & k_2 + k_3 & -k_3 \\ 0 & -k_3 & k_3 \end{bmatrix} \begin{Bmatrix} u_1 \\ u_2 \\ u_3 \end{Bmatrix} = \begin{Bmatrix} 0 \\ 0 \\ F_3(t) \end{Bmatrix}$$

which can be written as

$$[M]\{\ddot{u}\} + [K]\{u\} = \{P\}.$$

### 2.4.3  Obtaining Dynamic Characteristics of MDOF

We now have the forced response equations of motion. If we're just looking for the natural frequencies and modes shapes, we remove the forcing function and solve the eigenvalue problem (in turbopumps we will see that the loading pre-stresses components so significantly that their stiffness changes, so the loading can't be ignored for those problems.) We will seek solutions for the undamped, free vibration of the MDOF system with $N$ DOF's

$$[M]\{\ddot{u}\} + [K]\{u\} = \{0\}. \tag{2.13}$$

Assume a solution of the form

$$\{u\}_m = \{\phi\}_m e^{i(\omega_m t + \alpha_m)}$$

where $m = 1,\ldots,$ M (not the same M as the Mass Matrix), and M $\leq$ N (inequality to be discussed later). What's critically important to understand is that there are $m$ solutions, each with a different spatial part defined by $\phi$, which mathematically are the eigenvectors and which in vibration theory we call the mode shapes, and with each spatial solution there is an associated temporal solution with parameters of natural frequency and phase. This solution is completely analogous to the continuous solution. Figure 2.20 shows the relationship between the continuous and discrete modes, as well as the temporal (time history) solution.

$$\text{Continuous} \qquad \text{Discrete MDOF modes}$$
$$w(x) = U_m(x) \Leftrightarrow \{\phi\}_m$$

Examining the discrete MDOF section of the figure, we see that each degree of freedom will have its own spatial amplitude of $\phi_{im}$, where $i$ is the DOF and $m$ is the mode, as well as a time history. The time histories of each DOF show that they are completely in phase with each other. It's important to realize that for "normal" modes, which are derived without damping, all the DOF's are either completely in-phase or 180° out-of-phase with each other; the phase $\alpha$ is the relationship to the excitation of all the DOF's.

An example of multiple modes of a structure is shown in Fig. 2.21 for the cantilever beam, showing how the eigenvalues relate to the natural frequencies and eigenvectors. Animations of these mode shapes are also shown in Online Resource 1–3. The plus and minus signs indicate the phase relationship between various locations on the beam when vibrating in a particular mode. Another important characteristic shown in these modes is the existence of modal "nodes", which are locations where there is no motion throughout the cycle of vibration. This unfortunately is the same word used for the discretized locations of mass and stiffness in finite element analysis, so care must be taken with the context.

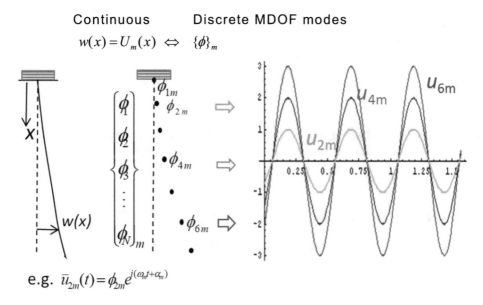

**Fig. 2.20** Comparison of continuous and discrete solution of a beam

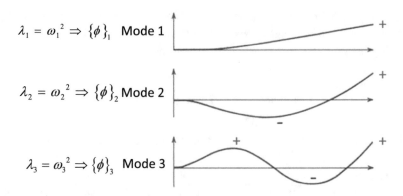

**Fig. 2.21** First three modes of cantilever beam

So, for example, if we have a 2 DOF system, the free vibration equations of motion in matrix form are

$$\begin{bmatrix} m_{11} & m_{12} \\ m_{21} & m_{22} \end{bmatrix} \begin{Bmatrix} \ddot{u}_1 \\ \ddot{u}_2 \end{Bmatrix} + \begin{bmatrix} k_{11} & k_{12} \\ k_{21} & k_{22} \end{bmatrix} \begin{Bmatrix} u_1 \\ u_2 \end{Bmatrix} = \begin{Bmatrix} 0 \\ 0 \end{Bmatrix}.$$

Examine the free vibration response of each DOF in the first mode:

$$u_{\text{dof1mode1}} = \phi_{11} e^{i(\omega_1 t + \alpha_1)}; \quad u_{\text{dof2mode1}} = \phi_{21} e^{i(\omega_1 t + \alpha_1)}$$

$$\ddot{u}_{11} = -\omega_1^2 \phi_{11} e^{i(\omega_1 t + \alpha_1)}; \quad \ddot{u}_{21} = -\omega_1^2 \phi_{21} e^{i(\omega_1 t + \alpha_1)}$$

Therefore,

$$e^{i(\omega_1 t + \alpha_1)} \left( [M] \begin{Bmatrix} -\omega_1^2 \phi_{11} \\ -\omega_1^2 \phi_{21} \end{Bmatrix} + [K] \begin{Bmatrix} \phi_{11} \\ \phi_{21} \end{Bmatrix} \right) = \begin{Bmatrix} 0 \\ 0 \end{Bmatrix}$$

$$-\omega_1^2 [M] \{\phi\}_1 + [K] \{\phi\}_1 = 0$$

Now we'll solve for $\omega$ and $\phi$ but generalize to the $m$'th mode instead of the first. This equation is now in the form of the "Generalized Eigenvalue Problem".

$$\left( [K] - \omega_m^2 [M] \right) \{\phi\}_m = 0.$$

Or we can simply name the matrix above as [D], the "System Matrix", where

$$[D] \{\phi\}_m = 0.$$

So for illustration, expand the above matrix equation:

$$\begin{bmatrix} k_{11} - \omega_m^2 m_{11} & k_{12} - \omega_m^2 m_{12} & \cdots & k_{1N} - \omega_m^2 m_{1N} \\ k_{21} - \omega_m^2 m_{21} & k_{22} - \omega_m^2 m_{22} & \cdots & k_{2N} - \omega_m^2 m_{2N} \\ \vdots & \vdots & \ddots & \vdots \\ k_{N1} - \omega_m^2 m_{N1} & k_{N2} - \omega_m^2 m_{N2} & \cdots & k_{NN} - \omega_m^2 m_{NN} \end{bmatrix} \{\phi\}_m = \{0\}$$

Implementing the above derivations into Part B of the procedure we started in the last section,

1. Write Equations of Motion and apply Boundary Conditions.
2. Write $[D]$; if masses "lumped", off-diagonals = 0.
3. Let $\lambda = \omega^2$.
4. Divide through by constant multiplier of $[K]$ and $[M]$ if available, and let:

$$\mu = \frac{\text{Constant multiplier of } [M]}{\text{Constant multiplier of } [K]} \lambda.$$

This step is not presented in other textbooks but is critical.
5. We now have $[D] = [k - \mu_{mm}]$ and $[D]\{\Phi\}_m = \{0\}$; and linear algebra says the determinant $|D| = 0$, which will give a polynomial equation of order N.

$$a_N \mu^N + a_{N-1} \mu^{N-1} + \ldots + a_1 \mu + a_0 = 0$$

6. Solve for roots $\mu_m$, i = 1, M modes; then solve for $\lambda = f(\mu)$, and $\omega = \sqrt{\lambda}$.
7. Plug $\mu_m$'s one at a time into $[D]\{\Phi\}_m = \{0\}$. Since D is singular (ie, "overdetermined"—too many unknown's), we have to make an assumption about value of one of the unknowns to solve for the others to determine $\{\Phi\}$. As previously mentioned, the actual magnitude of the modes is dependent upon a forcing function or initial condition. For this free vibration problem, we assign the magnitudes in two ways, either with "maximum normalization" or "mass normalization". For the first, we set the maximum value in the mode equal to one, and all the other values will be some percentage of it. This method is the most useful for basic understanding of the mode.
8. Mass normalization is required if you are going to use the modal matrix in additional mathematical calculations, i.e., "modal-superposition" forced response analysis. In this method, determine $\{\phi\}_m$ such that $\{\phi\}_m^T [M] \{\phi\}_m = 1$. This is accomplished by solving for a scale factor $s_m$ such that

$$s_m \{\phi\}_m^T [M] s_m \{\phi\}_m = 1$$

$$s_m = \frac{1}{\sqrt{\{\phi\}_m^T[M]\{\phi\}_m}}$$

The new "mass normalized" mode shapes are simply

$$\{\phi\}_m^{\text{mass}} = s_m\{\phi\}_m. \tag{2.14}$$

And finally, we frequently combine these modes on a column-wise base to form the "modal matrix" $\Phi$.

$$[\Phi] \equiv [\{\phi\}_1, \{\phi\}_2, \ldots, \{\phi\}_M].$$

In addition to being the mathematical basis for modal forced response analysis (to be discussed), the modes in of themselves are extremely valuable, particularly in liquid rocket engine structural dynamics. If there is a harmonic excitation, a resonant situation can occur, just like in SDOF systems. Another primary use (for all areas of structural dynamics) is matching analytical or numerical model natural frequencies and modes with those obtained from modal test (discussed in Chap. 7) to validate the model. Finally, many structural components and systems have requirements specifically identifying minimum frequency margins from fundamental or problematic modes.

### 2.4.4  Finite Element Example

Let's also go through an example of calculating the axial modes and natural frequencies of a finite element model of a cantilever bar to clarify some of the steps discussed. The bar is $960'' \times 2'' \times 1''$, with Young's Modulus E = 2.9e7 lb/in$^2$, mass density $\rho = 0.1 \frac{lb-sec^2}{in^3}$, and will be discretized into two finite elements, with the coordinate system as shown in Fig. 2.22.

Lump the mass at the nodes

$$m_{element} = \rho A L$$

So $m_1 = m_3 = \frac{\rho A L}{2}; \quad m_2 = \rho A L$
and the elements have stiffness

$$k_1 = k_2 = AE/L.$$

Now we follow the procedure, parts A and B:

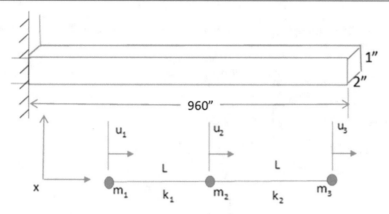

**Fig. 2.22** Cantilever Bar for Finite Element Analysis

$$k_1(u_2 - u_1) = m_1\ddot{u}_1$$
$$m_1\ddot{u}_1 - k_1(u_2 - u_1) = 0$$
$$m_1\ddot{u}_1 - k_1u_2 + k_1u_1 = 0$$

Similarly,

$$k_2(u_3 - u_2) - k_1(u_2 - u_1) = m_2\ddot{u}_2$$
$$m_2\ddot{u}_2 + k_2u_3 - k_1u_2 - k_2u_2 + k_1u_1 = 0$$
$$m_2\ddot{u}_2 + k_1u_1 + (-k_1 - k_2)u_2 + k_2u_3 = 0$$

and

$$k_2(u_3 - u_2) = m_3\ddot{u}_3$$
$$m_3\ddot{u}_3 - k_2(u_2 - u_3) = 0$$
$$m_3\ddot{u}_3 - k_2u_2 + k_2u_3 = 0.$$

The resulting equations of motion are shown below; in addition, we apply the boundary condition of $u_1 = 0$ by simply crossing out the first rows and columns of all the matrices.

$$\frac{\rho AL}{2}\begin{bmatrix} 1 & 0 & 0 \\ 0 & 2 & 0 \\ 0 & 0 & 1 \end{bmatrix}\begin{Bmatrix} \ddot{u}_1 \\ \ddot{u}_2 \\ \ddot{u}_3 \end{Bmatrix} + \frac{AE}{L}\begin{bmatrix} 1 & -1 & 0 \\ -1 & 2 & -1 \\ 0 & -1 & 1 \end{bmatrix}\begin{Bmatrix} u_1 \\ u_2 \\ u_3 \end{Bmatrix} = \begin{Bmatrix} 0 \\ 0 \\ 0 \end{Bmatrix}$$

Now write the system matrix $[D] = ([K] - \lambda^2[M])$, and let $\lambda = \omega^2$

$$[D] = \left(\frac{AE}{L}\begin{bmatrix} 2 & -1 \\ -1 & 1 \end{bmatrix} - \lambda\frac{\rho AL}{2}\begin{bmatrix} 2 & 0 \\ 0 & 1 \end{bmatrix}\right).$$

Now we divide through by the constant coefficients of M and K as shown below

$$\mu = \frac{\frac{\rho A L}{2}}{\frac{AE}{L}}\lambda = \frac{\rho L^2}{2E}\lambda$$

then $[D] = \begin{bmatrix} 2 - 2\mu & -1 \\ -1 & 1 - \mu \end{bmatrix}$.

Now set $\|[D]\|=0$

$$\begin{vmatrix} 2 - 2\mu & -1 \\ -1 & 1 - \mu \end{vmatrix} = 0$$

$$(2 - 2\mu)(1 - \mu) - 1 = 0$$

$$2\mu^2 - 4\mu + 1 = 0.$$

We now solve for the roots $\mu_m$, and then solve for $\omega_m$

$$\mu_{1,2} = \frac{4 \pm \sqrt{16 - (4 * 2 * 1)}}{2 * 2} = 1 \pm \frac{\sqrt{2}}{2}$$

$$\mu_1 = 0.2928 \Rightarrow \lambda_1 = \frac{2E}{\rho L^2}\mu = \frac{2(2.9e7\,\frac{lb}{in^2})}{0.1\frac{lb\,sec^2}{in^4}(480\,in)^2}0.2928 = 737.1$$

$$\Rightarrow \omega_1 = 27.15\,\frac{Rad}{sec}$$

$$\mu_2 = 1.707 \Rightarrow \lambda_2 = \frac{2(2.9e7\frac{lb}{in^2})}{0.1\frac{lb\,sec^2}{in^4}(480in)^2}1.707 = 4297.5$$

$$\Rightarrow \omega_2 = 65.5\,\frac{Rad}{sec}$$

The next step is to solve for the modes. First, plug $\mu_1$ into the first line of the matrix equation $[D]\{\phi\} = 0$, and solve for $\{\phi\}$ using "maximum" normalization:

$$\mu_1 : (1 - 0.2928)\phi_{11} - \phi_{21} = 0$$

$$0.7071\phi_{11} - \phi_{21} = 0 \Rightarrow \phi_{21} = 0.7071\phi_{11}$$

$$\text{Let } \phi_{11} = 1 \Rightarrow \phi_{21} = 0.7071 \Rightarrow \{\phi\}_1 = \begin{Bmatrix} 1 \\ 0.7071 \end{Bmatrix}$$

The second equation in the matrix equation will yield the same result, as it is not linearly independent from the first equation. Follow the same procedure to get the second mode:

$$\mu_2 : (1 - 1.7071)\phi_{12} - \phi_{22} = 0$$
$$- 0.7071\phi_{12} = \phi_{22}$$

$$\text{Let } \phi_{12} = 1 \Rightarrow \phi_{22} = -0.7071 \Rightarrow \{\phi\}_2 = \left\{ \begin{array}{c} 1 \\ -0.7071 \end{array} \right\}$$

Now concatenate the two modal vectors to form the Modal Matrix:

$$[\Phi] = \begin{bmatrix} 1 & 1 \\ 0.7071 & -0.7071 \end{bmatrix}$$

These max normalized mode shapes are useful for visualization. As shown in Fig. 2.23, in mode 1, both DOF's are in-phase with each other, while in mode 2, the DOF's are out-of-phase with each other. Again, it is critical to remember that the relative phase is different than the phase we obtain when solving forced response problems, which is the phase between the excitation and each DOF. Animations of some discrete MDOF systems are shown in an excellent presentation generated by the University of Pennsylvania, Online Resource 4 [5].

If forced response analysis is necessary, these modes can then be "mass" normalized. First calculate the scale factor $s_1$ and $s_2$ as described previously, which yields $s_1 = s_2 = 0.10206$, then apply Eq. (2.14) to generate the mass normalized modal matrix

$$[\Phi]_{mass} = \begin{bmatrix} 0.10206 & 0.10206 \\ 0.07216 & -0.07216 \end{bmatrix}.$$

### 2.4.5   Comparison with Closed-Form Solution

For this very simple system, a closed-form continuous solution using the handbook by Blevins can be used to identify the error of the numerical analysis. For this type of structure, the solution is

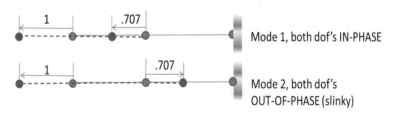

**Fig. 2.23**   Solution of finite element cantilever bar

$$f_i = \frac{\lambda_i}{2\pi L} \sqrt{\left(\frac{E}{\rho}\right)} \quad \text{where } \lambda_i = \frac{(2i-1)\pi}{2}$$

$$w(x) = \sin \frac{\pi(2i-1)x}{2L}$$

So plugging in the numerical parameters,

$$f_1 = \frac{(2(1)-1)\pi}{2(2)\pi 960} \sqrt{\frac{2.9 \times 10^7}{0.1}} = 4.43 \text{ Hz}$$

$$\omega_1 = f_1 2\pi = 27.86 \frac{\text{Rad}}{\text{sec}}$$

The numerical model generated $\omega_1 = 27.15 \frac{\text{Rad}}{\text{sec}}$, for an error of 2.6%. However, the exact value of the second natural frequency is $\omega_{2exact} = 83.593 \frac{\text{Rad}}{\text{sec}}$, compared to $\omega_{2numerical} = 65.5 \frac{\text{Rad}}{\text{sec}}$ for an error of 21.6%. The error of the first natural frequency is acceptable, but the second one is not. The large error is due to the method of discretizing the mass in the numerical model, which is to "lump" the masses at the DOF's where it is actually spread out. A more accurate technique called the "coupled" mass technique, which is discussed in finite element texts, will give much better answers and should be chosen if the number of DOF's in the finite element model is fairly limited; as the mesh becomes more and more refined, the lumped mass method will approach the correct answer (and it does run faster). On the other hand, the analytical mode shapes as defined in Blevins match the numerical method exactly:

$i = 1$ :

$$w(0.5L) = \sin \frac{\pi(2i-1)(0.5L)}{2L} = \sin\left(\frac{0.5\pi}{2}\right) = \sin \frac{\pi}{4} = 0.707$$

$$w(1L) = \sin \frac{\pi}{2} = 1$$

$i = 2$ :

$$w(0.5) = \sin\left(\frac{3\pi.5L}{2L}\right) = \sin \frac{1.5\pi}{2} = \sin \frac{3\pi}{4} = 0.707$$

$$w(1L) = \sin \frac{3\pi L}{2L} = \sin \frac{3\pi}{2} = -1$$

### 2.4.6 Free-Free and Rigid Body Modes

A strange type of "rigid-body-mode" comes about when a structure has no fixed DOF's on its boundaries, i.e., when it's in a "free-free" boundary condition state. Given the system and solution shown in Fig. 2.24.

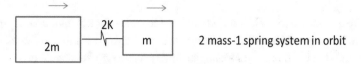

**Fig. 2.24** Free-free system

Drawing the Free Body Diagram to generate the equations of motion yields:

$$2k(u_2 - u_1)$$
$$2ku_2 - 2ku_1 = 2m\ddot{u}_1$$

$$2k(u_2 - u_1)$$
$$-(2ku_2 - 2ku_1) = m\ddot{u}_1$$

$$\begin{bmatrix} 2m & 0 \\ 0 & m \end{bmatrix} \begin{Bmatrix} \ddot{u}_1 \\ \ddot{u}_2 \end{Bmatrix} + \begin{bmatrix} 2k & -2k \\ -2k & 2k \end{bmatrix} \begin{Bmatrix} u_1 \\ u_2 \end{Bmatrix} = \begin{Bmatrix} 0 \\ 0 \end{Bmatrix}$$

The characteristic equation is therefore

$$m\lambda(2m\lambda - 6k) = 0$$

which has roots:

$\lambda_{1,2} = 0, \frac{3k}{m}$. Now, plug in $\lambda_1 = 0$ to get $\{\phi\}_1$

$$\begin{pmatrix} 2k & -2k \\ -2k & 2k \end{pmatrix} \{\phi\}_1 = 0$$

Solving for $\{\phi\}$,

$2k\phi_{11} = 2k\phi_{12} \Rightarrow \{\phi\}_1 = \begin{Bmatrix} 1 \\ 1 \end{Bmatrix}$, i.e., both bodies move in-phase together as a

rigid-body with a natural frequency of 0. If we then plug in $\lambda_2 = \frac{3k}{m}, \Rightarrow \{\phi\}_2 = \begin{Bmatrix} 1 \\ -2 \end{Bmatrix}$,

called the first "flexible mode", then the modal matrix is $[\Phi] = \begin{pmatrix} 1 & 1 \\ 1 & -2 \end{pmatrix}$. An animation

of the first free-free flexible body mode is shown in Online Resource 5. This type of model has a variety of extremely valuable applications. First, of course, it is a representative boundary condition for complete models of aerospace vehicles. In addition, it is the easiest numerical boundary condition to replicate in test, and so it is frequently used for model validation via modal test. This is accomplished by placing the object on a compliant surface with its own natural frequency less than 1/10 of the first "flexible" (non-zero) frequency of the structure, e.g., a foam pad or hanging from bungee cords. Finally, this concept is valuable for finite element model verification. The number of rigid-body close-to-zero-frequency modes will equal the number of degrees of freedom per node in a

FE model (e.g., 6 for beam-element models, 3 for solid-element models). If the analysis yields more, it is a good indication that there is a mechanism or hinge in the model, which is probably unintended due to a modeling error like a gap or disconnected mesh interfaces.

### 2.4.7  Cyclically Symmetric Structural Modal Properties

One of the most important aspects of modal analysis of LRE's is that many of the components possess the property of cyclic symmetry, in which a structure is composed of sectors that are identical when rotated about a central axis. The sectors do not have to be perfect "pie" wedges with straight sides, i.e., they can be curved. An example is shown in Fig. 2.25. This characteristic is vitally important because spatial Fourier decomposition techniques have been developed such that only one sector needs to be modelled, and the results for the other sectors can be determined using closed-form equations that are built into FE software packages. As the size of the finite element matrices is proportional to the square of the number of DOF's, the cyclic symmetry representation therefore saves enormously on computing costs. In fact, analysis of bladed-turbine disks, which can have over 100 blades, each on an identical sector, is only possible using the cyclic symmetric analysis methodology.

The theoretical basis for this methodology is not trivial and will not be presented here, but the reader is referred to Christensen [6] for these details. We will, however, discuss several of the relevant characteristics of cyclic symmetric modes, and these characteristics will be referred to frequently in later sections of this text. The first of these is that the modes are classified into families defined by the number of waves either in the axial, circumferential, or radial directions. This verbal classification is critical in structural dynamic analysis of LRE's, unlike other structural dynamic analysis, because true resonant response is a function of these classifications.

The lowest frequency modes are generally those with waves only in the radial direction, and every sector has the exact same shape (see Fig. 2.26). The remainder of the modes are generally composed of waves in the circumferential direction (although they

**Fig. 2.25**  Finite element model of turbopump impeller showing curved cyclic sector

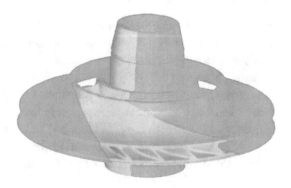

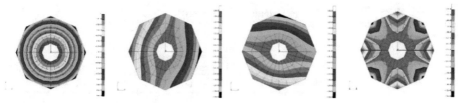

**Fig. 2.26** Families of cyclic symmetric mode shapes

can occur alongside waves in the other coordinates), and since the nodes of this mode shape lie along diametral lines of the structures, these modes are generally referred to as "nodal diameter" modes. The lower modes with only radial waves (or axial waves) are designated as "0" nodal diameter modes.

One fascinating aspect of the middle set of nodal diameter modes is that they occur in identical, or "degenerate" pairs of eigenvalues, with mode shapes identical but differing by an angle equal to half the angle of a circumferential wave. For example, in the 1 nodal diameter mode shown in the figure, each half wave is 180°, so half of that is 90°, resulting in the two modes, with identical natural frequencies but rotated shapes. In real structures, this will never occur, because each sector can never be truly identical due to manufacturing variations (called mistuning), so the pair will differ slightly. Finally, for an even number of sectors, the last nodal diameter type mode will consist of a single mode with a unique eigenvalue.

The next critical aspect of cyclic symmetry analysis is the setup of the Fourier parameters in the analysis. FE codes require that the user input a "harmonic" which identifies which nodal-diameter families will be obtained using the methodology. In general, a user will only need a subset of these families since they are the only ones that will be responsive to the excitation shape (to be discussed in later sections). The relationship of the ND families to the harmonics follows Eq. 2.15.

$$ND = qS \pm k, \text{ where } q = 0, 1, 2, \ldots \tag{2.15}$$

where $S$ is the number of sectors and $k$ is the harmonic. For example, if there are an even number of sectors, such as the example of 8 in Fig. 2.23, then the following relationship will exist, following a stair-step type of pattern, as shown in Table 2.1. The equation holds

**Table 2.1** Nodal diameters as function of harmonic family for $N = 8$

| Harmonic | ND = f(q) | ND | | | | ... |
|----------|-----------|-----|-----|-----|-----|-----|
| 0 | 8q | 0 | 8 | | 16 | |
| 1 | 8q ± 1 | 1 | 7 | 9 | 15 | 18 |
| 2 | 8q ± 2 | 2 | 6 | 10 | 14 | 19 |
| 3 | 8q ± 3 | 3 | 5 | 11 | 13 | 20 |
| 4 | 8q ± 4 | 4 | | 12 | | 21 |

**Table 2.2** Nodal diameters as function of harmonic family for N = 11

| Harmonic | ND = f(q) | ND | | | | |
|---|---|---|---|---|---|---|
| 0 | 11q | 0 | 11 | | 22 | |
| 1 | 11q ± 1 | 1 | 10 | 12 | 21 | 23 |
| 2 | 11q ± 2 | 2 | 9 | 13 | 20 | 24 |
| 3 | 11q ± 3 | 3 | 8 | 14 | 19 | 25 |
| 4 | 11q ± 4 | 4 | 7 | 15 | 18 | 26 |
| 5 | 11q ± 5 | 5 | 6 | 16 | 17 | 27 |

for an odd number of sectors, although the stair-step pattern is a little different for the (N − 1)/2 harmonic family, as shown in Table 2.2 for N = 11.

For many structures, a given mode will have components of a number of nodal diameters within a harmonic family. For example, the 6 sector SSME Fuel Turbopump 3rd stage impeller (Fig. 2.26), harmonic 2, shows strong presence of all the first 4 nodal diameters (2, 4, 8, 10).

## 2.5  Forced Response of Damped MDOF Systems

We now move into forced response of damped multi-DOF systems, which is closest to the general condition of structures in dynamic environments. There are three kinds of forced response analysis:

1. Frequency Response (also called Harmonic Response), which consists of an excitation decomposed into the frequency domain, and each of those components is applied separately to the structure for which the response at individual frequencies is calculated.
2. Transient Response, consisting of an explicitly defined, complete time history of an excitation applied to the structure in the time domain, and the response is calculated in the time domain. As mentioned previously in Sect. 2.2.3, it is called "transient" because the solution is calculated using both the steady-state and transient parts of the solution to the differential equations of motion, not because it fades away.
3. Random response, consisting of non-periodic excitation that is not explicitly defined in time, but is defined continuously in the frequency domain, and the response is calculated also in the frequency domain.

A forced response analysis is a non-trivial exercise, even with the huge advances in numerical techniques, and still has significant uncertainty generally due to large uncertainty in the forcing function and damping. The following procedure should therefore be followed when evaluating a structure under dynamic loading. First, a "quasi-static" analysis should be performed, in which the dynamic load is applied statically. This is an excellent "sanity" check; as the non-resonant behavior of the structure (which is not dependent on damping) should be of the same magnitude as this result, the analysis is generally simpler and less prone to assumptions in the excitation, and it can also be used for comparison to resonant response calculations. A quick modal analysis obtaining just the first few modes should be performed next, making sure to use operating boundary conditions (i.e., not modal test boundary conditions, which are frequently non-operational, to be discussed in Chap. 7). If the excitation frequency range is less than a third of the first (also called fundamental) natural frequency, the quasi-static results can be used for the structural assessment, including not only ultimate or yield, but also fatigue capability or displacement requirements. Finally, if the excitation is greater than a third of the fundamental natural frequency, a more thorough modal analysis and modal test, if possible, should be performed to verify that the mode is well separated from the excitation frequency, which would be a resonant condition. If resonance doesn't exist, the quasi-static results can still be used, and if it does, the complete forced response analysis must be performed.

### 2.5.1 Frequency (Harmonic) Response Analysis

A frequency response is the calculation of the steady-state solution $u(\Omega)$ for a given, constant amplitude harmonic excitation. Any quantity of interest can be obtained, e.g., displacement, stress, or acceleration. Even if the excitation is not substantially harmonic (although many excitations in LRE's are), the information obtained by a frequency response is extremely helpful in evaluating the importance of different modes and different excitation frequency ranges, which could be tied to different conditions. This section is quite laden with equations but is quite important. In particular, the last step, which is necessary to actually get a usable equation that can implemented to calculate a solution, is usually skipped in SD textbooks.

### 2.5.2 Equations of Motion Using Modal Superposition

Central to a frequency response analysis is the concept of Modal Superposition and Modal (or Generalized) Coordinates. Starting with Eq. (2.12), we add a damping coefficient matrix for the velocity vector, and a forcing vector $\{P(t)\}$

$$[M]\{\ddot{u}\} + [C]\{\dot{u}\} + [K]\{u\} = \{P(t)\} \tag{2.16}$$

This matrix equation represents a series of "coupled" ordinary differential equations, where coupled means that the individual equations will not be in terms of a single DOF but will instead have terms involving multiple DOF's. We've already seen this in our modal solution of the multiple DOF system in Sect. 2.4. This coupling means we can't use the closed-form solution for a SDOF system derived in Sect. 2.3 and must instead use an iterative numerical scheme. While this is certainly possible, it is prohibitively expensive for a finite element model with more than a few thousand DOF's, which virtually all models have (many are in the millions of DOF's). Expensive in this context means computer time—weeks for a solution.

A fantastically elegant technique called the modal superposition technique has therefore been developed which solves this issue. This technique transforms the coupled system to a set of uncoupled, SDOF equations which can be solved with the SDOF methods. The first step in the process is to obtain the mass-normalized modal matrix $[\Phi]_{\text{mass}}$ which is of size N by M, as denoted. Then introduce a coordinate transformation from the "physical coordinates $\{u\}$ to a set of "generalized", also called "modal" coordinates $\eta$ of size M", which are at this point just a convenient mathematical construct:

$$\{u\} = N \overset{M}{[\Phi]}\{\eta\}M \tag{2.17}$$

which we plug into Eq. (2.16) to give

$$[M][\Phi]\{\ddot{\eta}\} + [C][\Phi]\{\dot{\eta}\} + [K][\Phi]\{\eta\} = \{P(t)\}. \tag{2.18}$$

Now pre-multiply both sides by $[\Phi]^T$

$$[\Phi]^T[M][\Phi]\{\ddot{\eta}\} + [\Phi]^T[C][\Phi]\{\dot{\eta}\} + [\Phi]^T[K][\Phi]\{\eta\} = [\Phi]^T\{P(t)\}$$

Recall that if the modal matrix is mass normalized, then by definition $[\Phi]^T[M][\Phi] = [^{\backslash}I_{\backslash}]$. In addition, a mathematical property called orthogonality states that given a set of eigenvectors, then

$$\{\phi\}_i[K]\{\phi\}_j = \{0\}. \tag{2.19}$$

It then follows, therefore, that

$$[\Phi]^T[K][\Phi] = [^{\backslash}\kappa_{\backslash}]$$

where $[^{\backslash}\kappa_{\backslash}]$ is only defined (at this point) to be a diagonal matrix.

In addition, if we make the convenient assumption of "proportional damping", which is that the damping coefficient matrix can be expressed as a linear combination of the mass and stiffness matrix (this assumption may appear to be arbitrary and without physical justification, but it is actually not too bad)

$$[C] = \alpha[M] + \beta[K]$$

then it turns out [C] will also satisfy orthogonality, i.e.,

$$[\Phi]^T[C][\Phi] = [^{\backslash}C_{\backslash}],$$

where $\lceil^\backslash C\backslash\rceil$ is only defined (at this point) to be a diagonal matrix, too. Putting it all together, we therefore now wind up with a completely uncoupled (all the coefficient matrices are diagonal) series of equations in $\eta$

$$\lceil^\backslash I\backslash\rceil\{\ddot{\eta}\} + \lceil^\backslash C\backslash\rceil\{\dot{\eta}\} + \lceil^\backslash \kappa\backslash\rceil\{\eta\} = [\Phi]^T\{P(t)\}. \tag{2.20}$$

Now going back to the section on SDOF forced response, we restate the SDOF equation of motion, Eq. (2.8)

$$\ddot{u} + 2\varsigma\omega\dot{u} + \omega^2 u = F\cos\Omega t.$$

Comparing this with the individual equations comprising Eq. (2.20), we see that they are of the same form, where $\eta$ takes the place of $u$. Therefore, each value of $\lceil^\backslash C\backslash\rceil$ must equal $2\zeta_i\omega_i$, where the $\zeta_i$ are defined to be "modal damping," which can be obtained in modal test or estimated based upon experience, and each value of $\lceil^\backslash \kappa\backslash\rceil$ must equal $\omega_i^2$, which are the natural frequencies. Putting these together, we get $M$ single degree of freedom equations of motion in $\eta$, with $m = 1, \ldots, M$:

$$\ddot{\eta}_m + 2\zeta_m\omega_m\dot{\eta}_m + \lambda_m\eta_m = \{\phi\}_m^T\{P(t)\} \tag{2.21}$$

### 2.5.3   Solution of MDOF Frequency Response Using Modal Superposition

Let $\{p(t)\} = \{F\}e^{i\Omega t}$, i.e., a time varying forcing function with different amplitudes at different $n$ DOF's, but all harmonics at the same excitation frequency $\Omega$ and in phase with each other. Now let's calculate the response for the $m$'th generalized coordinate $\eta_m$ using the SDOF solution from Sect. 2.3, restating Eq. (2.10) into

$$u(t) = \frac{F_o}{K}\left|\overline{H}(\Omega)\right|e^{i(\Omega t + \phi)}$$

where

$$\left|\overline{H}(\Omega)\right| = \frac{1}{\sqrt{(1 - r^2)^2 + (2\zeta r)^2}}.$$

So the analogous solution for the "modal" coordinate $\eta$ is

$$\eta_m(t) = \frac{\{\phi\}_m^T\{F\}}{\lambda_m}\frac{1}{\sqrt{(1 - r_m^2)^2 + (2\zeta_m r_m)^2}}e^{i(\Omega t + \alpha_m)}$$

where

$$r_m = \frac{\Omega}{\omega_m} \text{ and } \alpha_m = ArcTan[(1 - r_m^2), (-2\zeta_m r_m)]$$

The phase is bit tricky; the correct complex quadrant of the numerator and denominator must be maintained, as indicated here by using the Mathematica© software code's notation of ArcTan.

But this solution is for the modal coordinates and although they do have some value in and of themselves, what we really want are the physical coordinates, which we obtain by substituting our solution for $\eta$ into the original transformation $\{u\} = [\Phi]\{\eta\}$, yielding

$$u_n(t) = \begin{bmatrix} \phi_{n1} & \phi_{n2} & \cdots & \phi_{nM} \end{bmatrix} \left\{ \left( \frac{\{\phi\}_m^T \{F\}}{\lambda_m} \left| \overline{H}_m(\Omega) \right| e^{i(\Omega\, t + \alpha_m)} \right) \right\}.$$

This is a harmonic solution, and the magnitude is obtained by setting the temporal part $e^{i(\Omega\, t + \alpha_m)} = 1$.

If this equation is written out, it is a series composed of coefficients of the terms $e^{i(\Omega\, t + \alpha_1)}$, $e^{i(\Omega\, t + \alpha_2)}$, , ..., which are impossible for matrices to handle in that polar (magnitude and phase) form of $\overline{H}_m(\Omega)$. However, if the real/imaginary form of these complex terms are used instead, where

$$\overline{H}_m(\Omega) = \frac{1 - r_m^2}{\left(1 - r_m^2\right)^2 + (2\zeta_m r_m)^2} + i\left( \frac{-2\zeta_m r_m}{\left(1 - r_m^2\right)^2 + (2\zeta_m r_m)^2} \right),$$

then our final complex solution can be written in a form that can be calculated (this step is the one ignored in most textbooks).

$$\overline{u}_n(t) = \left[ \phi_{n1} \frac{\{\phi\}_1^T \{F\}}{\lambda_1} \left( \frac{1 - r_1^2}{\left(1 - r_1^2\right)^2 + (2\zeta_1 r_1)^2} + i\left( \frac{-2\zeta_1 r_1}{\left(1 - r_1^2\right)^2 + (2\zeta_1 r_1)^2} \right) \right) \right.$$

$$+ \phi_{n2} \frac{\{\phi\}_2^T \{F\}}{\lambda_2} \left( \frac{1 - r_2^2}{\left(1 - r_2^2\right)^2 + (2\zeta_2 r_2)^2} + i\left( \frac{-2\zeta_2 r_2}{\left(1 - r_2^2\right)^2 + (2\zeta_2 r_2)^2} \right) \right)$$

$$\vdots$$

$$\left. + \phi_{nM} \frac{\{\phi\}_M^T \{F\}}{\lambda_M} \left( \frac{1 - r_M^2}{\left(1 - r_M^2\right)^2 + (2\zeta_M r_M)^2} + i\left( \frac{-2\zeta_M r_M}{\left(1 - r_M^2\right)^2 + (2\zeta_M r_M)^2} \right) \right) \right] e^{i\Omega\, t}$$

The real parts and the imaginary parts can now be added, i.e., standard matrix operations can be applied, and then the total real and imaginary converted back to the polar form

$$u_n(t) = |\bar{u}_n(t)| = A e^{i(\Omega t + \alpha)}$$

where $A$ is the magnitude of the complex response and $\alpha$ is the argument. So at the end of the day, we wind up with a simple equation expressed as

$$\{\bar{u}(t)\} = [\Phi]\left\{ \frac{\{\phi\}_m^T \{F\}}{\lambda_m} \bar{H}_m(\Omega) \right\} e^{i\Omega t}. \tag{2.22}$$

A short example of how the results are used is one from rocket engine turbopumps (to be expanded upon in Chap. 3). During the 1990's, the first stage blade in the Space Shuttle Main Engine turbopump developed a crack (Fig. 2.27) [7]. All the standard sources of excitation had been evaluated, but we realized that the gap between the seals in the housing around the blade tips would also cause a periodic excitation as the blade passed in-board of it, and this excitation had not been assessed (Fig. 2.28). A critical part of the blade assessment was determining the relative response of the blade modes to various excitation frequencies, so a MDOF frequency response as described above was calculated using a unit amplitude loading from the gap, with the results shown in Fig. 2.29. This analytical plot was critical in identifying the most responsive modes to the tip seal gap loading and to potentially reconfigure the number of gaps to reduce resonant conditions. This analysis was performed before Computation Fluid Dynamics could generate "actual" forcing functions in a reasonable time but shows that structural dynamic response analysis can still be beneficial without exact knowledge of the forcing function, which is still a source of significant uncertainty.

**Fig. 2.27** SSME turbine cross-section

1st Blade

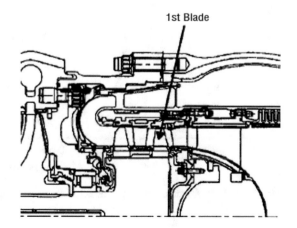

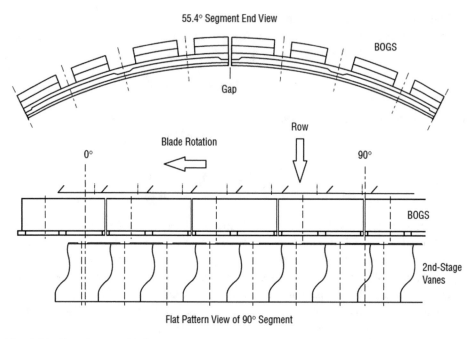

**Fig. 2.28** Gaps between blade outer gas seals

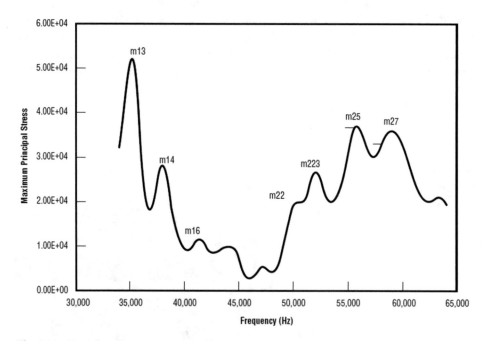

**Fig. 2.29** Total frequency response function showing modal response

## 2.5.4 Non-resonant Forced Response

It is sometimes easy to forget that structures will respond to harmonic forced excitation at the frequency of that excitation, even if its frequency is not close to a natural frequency. This response is called "non-resonant forced response" and must be evaluated whenever there is a dynamic forcing function. It is also tempting to simply assume that this response will be the same as the static response, which is the case for sub-resonant excitation of a SDOF system, but as the curve in Fig. 2.29 shows, the response in between modes for a structure with high modal density is not simply this single value. For these high modal density structures, the complete frequency response curve must be used. As a result, now that finite element codes have frequency response analyses based upon the MDOF curves that are easy to perform, it is generally worthwhile to do those analyses even for these non-resonant cases rather than make a wildly inaccurate estimate based upon an inappropriate SDOF assumption. Occasionally, though, it is clear that the excitation is significantly sub-resonant as mentioned at the beginning of this section, so the static response is a good "back-of-the-envelope" estimate for those cases.

## 2.5.5 Modal Truncation

In addition to decoupling the equations of motion and thereby enabling a closed-form solution of the forced response, the use of modal superposition enables the structural dynamic system to be represented by an order-of-magnitude smaller subset of the modes. Up to this time the modal matrix $[\Phi]$ has been a square matrix of order $N$ degrees of freedom. An approximate solution can be obtained, though, by truncating the number of modes, which are the columns, in $[\Phi]$ to some value $M$, i.e., $N[\overset{M}{\Phi}]$, where $M$ modes $\ll N$ DOF's. We can then use this adapted modal matrix in the transformation equation 2.16, repeated here, to the modal coordinates $\eta$

$$N\{u\} = N[\overset{M}{\Phi}]\{\eta\}M \tag{2.23}$$

This transformation to a reduced modal set is physically intuitive. Almost all structural dynamic problems have some finite frequency range of interest, both in the excitation and/or the desired response, and these ranges are used to determine the natural frequencies of the truncated set of modes used in Eq. 2.23. Because the solution using this truncation is not the exact answer, though, the general guideline is to use a frequency range twice as large as the range of interest.

## 2.5.6    Modal Force

The modal (also called generalized) force is defined as the right half side of the modal forced response equation of motion, Eq. (2.22).

$$\{\mathcal{F}\}_m = \{\Phi\}_m^T\{P\}.$$

It is equivalent to the dot product of each mode with the excitation vector, and is a vector of size $M$, one element for each mode. Physically it expresses the similarity of the spatial shape of each mode with the spatial shape of the force. Generally, the larger the modal force, the larger the response of a mode.

For example, look at the 3-DOF problem in Fig. 2.30.

Using the procedure discussed in Sect. 2.4.3 (an excellent exercise for the reader), the mass normalized modal matrix is

$$[\Phi] = \begin{bmatrix} 0.2761 & 0.6667 & 0.6923 \\ 0.3983 & 0.3333 & -0.4798 \\ 0.4496 & -0.3333 & 0.1417 \end{bmatrix}$$

Let's say the force vector is shaped exactly like mode 1, i.e.

$$\{P\} = \{\phi\}_1 = \begin{Bmatrix} 0.2761 \\ 0.3983 \\ 0.4496 \end{Bmatrix}.$$

What is the modal force resulting from the interaction with mode 3?
Using the definition, we get

$$\mathfrak{I} = \{\phi\}_3^T\{P\} = [0.6923 \quad -0.4798 \quad 0.1417] \begin{Bmatrix} 0.2761 \\ 0.3983 \\ 0.4496 \end{Bmatrix}$$

$$= (0.1911 - 0.1911 + 0.0637) = 0.0637$$

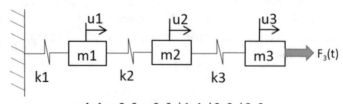

m1=1, m2=2, m3=3, k1=1, k2=2, k3=3

**Fig. 2.30**  3-DOF system

Compare this result with applying a force vector that is the same shape as mode 3:

$$\Im = \{\phi\}_3^T \{P\} = [0.6923 \quad -0.4798 \quad 0.1417] \begin{Bmatrix} 0.6923 \\ -0.4798 \\ 0.1417 \end{Bmatrix}$$

$$= 0.7296$$

which is 11.5 times larger. In the limit as the DOF approaches infinity, the modal force of a force vector shaped like one mode applied to a different mode will be zero since structural eigenvectors are orthogonal. But even with just 3 DOF, this result illustrates that the most effective way to distribute a load to excite a mode is to match the load with the mode shape.

### 2.5.7   Transient Response of MDOF Systems

The "transient" response is defined to be the temporal (time-domain) response due to an arbitrary excitation explicitly (and usually discreetly) defined in time for each degree of freedom (see Fig. 2.31).

There are two main solution techniques for calculating the response. The first is analytically using the impulse response function, which is only practical for small problems. Details on that technique can be found in the text by Craig [8]. The second method is numerical and is a "time-marching" iterative procedure that solves for the solution at every time step as a function of the system state at the previous and future time steps, where the future state is the unknown. This method in turn can be solved either using either modal superposition, as described in the last section, to reduce the system equations of motion into uncoupled equations of motion, or directly, in which the matrix equations are solved numerically. Although the practitioner won't be programming the solution, it's valuable to see the basic technique to help to understand its applicability, and so is described here (also from Craig) [9].

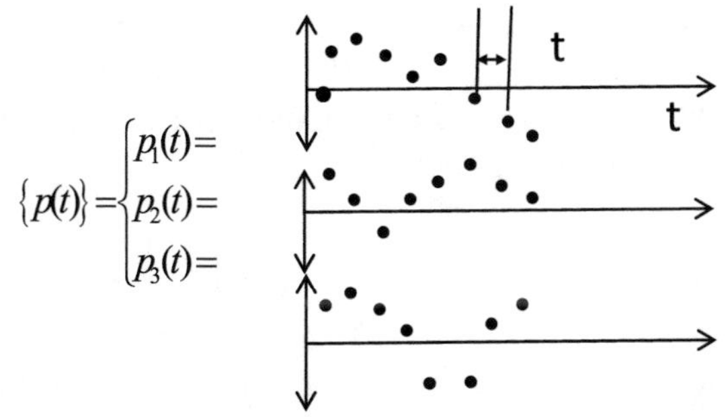

**Fig. 2.31** "Transient" forcing function

$$\{p(t)\} = \begin{Bmatrix} p_1(t) = \\ p_2(t) = \\ p_3(t) = \end{Bmatrix}$$

Given the MDOF forced vibration equations of motion

$$[M]\{\ddot{u}(t)\} + [C]\{\dot{u}(t)\} + [K]\{u(t)\} = \{P(t)\}$$

define the velocity and acceleration vectors in terms of the displacement vector at the time step immediately preceding and following the present step, where $n$ is the time increment.

$$\{\dot{u}_n\} = \frac{1}{2\Delta t}\{u_{n+1} - u_{n-1}\}$$

$$\{\ddot{u}_n\} = \frac{1}{\Delta t^2}\{u_{n+1} - 2u_n + u_{n-1}\}$$

If the force and displacements are averaged over the three time points, the equation of motion becomes

$$\left[\frac{M}{\Delta t^2}\right](u_{n+1} - 2u_n + u_{n-1}) + \left[\frac{C}{2\Delta t}\right](u_{n+1} - u_{n-1}) + \left[\frac{K}{3}\right](u_{n+1} + u_n + u_{n-1})$$
$$= \frac{1}{3}(P_{n+1} + P_n + P_{n-1})$$

Solving for $\{u_{n+1}\}$:

$$\{u_{n+1}\} = \frac{\frac{1}{3}\{P_{n+1} + P_n + P_{n-1}\} + \left[\frac{2M}{\Delta t^2} - \frac{K}{3}\right]\{u_n\} + \left[-\frac{M}{\Delta t^2} + \frac{C}{2\Delta t} - \frac{K}{3}\right]\{u_{n-1}\}}{\left[\frac{M}{\Delta t^2} + \frac{C}{2\Delta t} + \frac{K}{3}\right]}$$

### 2.5.8   Base Excitation of MDOF Systems

Implementation of base acceleration excitation (also known as enforced acceleration) for MDOF systems introduces several complications over the SDOF equations shown in Sect. 2.3.4. This technique is the basis of random vibration analysis for most components of LRE's outside of turbopumps. Although these procedures are performed within finite element codes, it is worth going through a brief derivation for more complete understanding as well as to clearly see how to handle the DOF with the specified excitation.

As shown by Paultre [10], the relative (to the base) displacement vector can be partitioned into the DOF whose displacement (or acceleration) is being specified $u_2$ (or $\ddot{u}_2$) and the rest of DOF's $\{u_1\}$

$$\{u\} = \left\{\begin{array}{c} u_1 \\ u_2 \end{array}\right\} = \left\{\begin{array}{c} u_1^{qs} \\ u_2 \end{array}\right\} + \left\{\begin{array}{c} y \\ 0 \end{array}\right\}$$

where $\{u^{qs}\}$ is the quasi-static portion of the response, that is the displacement resulting just from very low frequency excitation in which the inertia doesn't play a role, and $\{y\}$, the portion of the response due to the inertial dynamics. If this is substituted into the MDOF equation of motion, a couple of steps results in

$$\{u^{qs}\} = -[K_{11}]^{-1}[K_{12}]\{u_2\}$$

And the equation for the dynamic response $\{y\}$ becomes

$$[M_{11}]\{\ddot{y}\} + [C_{11}]\{\dot{y}\} + [K_{11}]\{y\} = [M_{11}][K_{11}][K_{12}]\{\ddot{u}_2\}$$

The first step in the analysis will be to perform a modal extraction, in which all the base DOF's are fixed. These modes can be used in modal superposition in the following forced response (harmonic, transient, or random) to solve for $\{y\}$ where that DOF is no longer fixed but its acceleration is defined for use in the equation showed above.

An alternative technique called the "Large Mass Method" has been used for many years. In this method, a huge mass, several orders of magnitude times the size of the structure being analyzed, is multiplied by the specified acceleration to generate a force which can then be used directly in the standard applied force equations of motion presented in Sect. 2.5.3. This method does involve some approximations, though, and has generally been superseded by the method described above.

### 2.5.8.1   Modal Participation Factor/Effective Mass

As previously mentioned, all modes are not created equal with regards to their effect on the dynamic response of a structure. A calculation can be made for base excited systems that quantifies this effect called the modal participation factor. This term is not consistently used in the literature, but an excellent breakdown of the different interpretations is presented in a review by Nieto [11]. The participation factor $P_i$ for mode $i$ is defined as [12]

$$P_i = [\phi_i^T][M]\{D_R\}$$

where $\{D_R\}$ is a rigid body vector in direction R. For example, for a two-element finite element model composed only of solid elements, which have DOF's in the X, Y, and Z directions, then

$$\{D_x\} = \begin{bmatrix} 1 & 0 & 0 & 1 & 0 & 0 \end{bmatrix}^T$$

The participation factor gives an idea of how important that mode is relative to the other modes and is important for determining which modes to focus on in model test/analysis validation, redesign to achieve particular structural dynamic objectives, and other uses.

It can also be shown that the square of $P_i$ is the "effective mass" of that mode, which when summed together for all modes equals the total mass of the structure. Generally, one would like to include enough modes so that this value is 85% of the total mass, implying that modal set chosen represents the structural dynamics adequately. An example of this calculation as applied to LRE nozzles will be presented in Sect. 4.4.2.

### 2.5.9  Component Mode Synthesis (CMS)

CMS is a very widely-used method closely related to modal superposition that is used to dynamically couple together substructures built by different organizations, or substructures that are modular. The basic premise is that each substructure is partitioned into internal DOF's and boundary DOF's, and the internal DOF's are transformed into reduced sets of substructure fixed-base modes, following the modal superposition techniques. The boundary DOF's are then coupled together using kinetic and potential energy equivalences. This allows modal tests and/or modal analysis of the components to be directly linked together into a system model, in addition to other benefits.

As liquid rocket engines are generally built by the same organization, and not many of the components are modular, it is rarely used in this application apart from the advanced technique of mistuning analysis of turbine bladed-disks. As it is not essential for general structural dynamic analysis of liquid rocket engines, and the derivation is somewhat lengthy, we will not cover it here, therefore. The reader is referred again to the text by Craig (who is a co-creator of the most widely used of the CMS methods, the Craig-Bampton/Hurty Method).

## 2.6  Random Vibration

For most of the structures in liquid rocket engines outside of turbopump flow path components, the primary component of the load is from sources generating random vibration. Random vibration is when the magnitude is never fully known at any point in time but instead can only be characterized statistically. If a structure undergoes random loading, its response can only be characterized statistically as well. The main source of random vibration in LRE's is the main combustion chamber, but pre-burners and gas generators also generate significant random forcing functions. Calculating the response to these loads requires combining frequency response techniques with statistics. The discipline is enormous, and entire textbooks and research conferences are devoted to just random vibrations, but we will cover the basics here considering its importance to LRE's.

As an example, assume you are in a vehicle driving along a bumpy road. The time history of the base excitation will be a quantity that can be repeatedly quantified statistically, but the value at a given point in time will unlikely repeat and cannot be predicted. To characterize the response, a random vibration analysis is performed based upon the following steps. First, a sample set of time histories of repeated drives along the stretch, called an ensemble, are taken as shown in Fig. 2.32 (the statistics of the condition need to stay constant, which is called a stationary process). Power spectral density (mathematically just called "spectral density") calculations using Eq. 2.22 are made on longer and longer histories, or more and more combined sets, until the value converges, as shown in Fig. 2.33.

$$PSD(\omega) = \frac{1}{\pi} \int\limits_{0}^{\infty} f(\tau)e^{i\omega\,t}d\tau \qquad (2.24)$$

The PSD describes the "spectral content" of the series of time histories, i.e., which harmonics make up the signal, and is more accurate as the time history gets longer, especially for low frequencies whose wavelength approach the length of the entire history. Random response analysis generally examines a set of PSD's of some sort of input, such as force, pressure, or base excitation in acceleration units, and generates the response to this input. Since a set of measured PSD's from a test generally vary somewhat, the actual

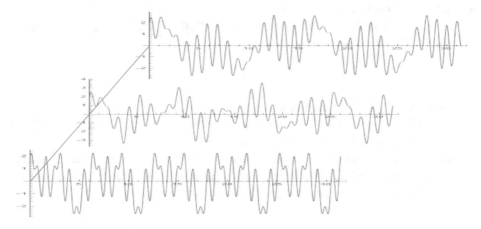

**Fig. 2.32** Population of temporal samples of a dynamic excitation or response

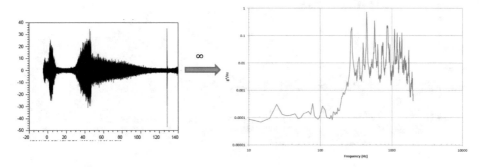

**Fig. 2.33** Transformation of Ensemble time histories to PSD

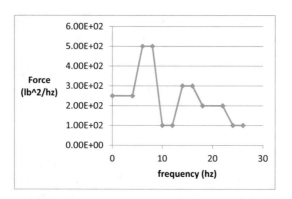

**Fig. 2.34** Envelope of PSD's

input used for the analysis, $S_{xx}(\Omega)$, is generated by enveloping the set, as shown in Fig. 2.34. This adds significant conservatism to the environment, which can be seen by comparing the RMS of the envelope to individual PSD samples.

The Complex Frequency Response $H(\Omega)$ can then be used to directly calculate a response $S_{yy}(\Omega)$ to this forcing function.

$$S_{yy}(\Omega) = H(\Omega)S_{xx}(\Omega)$$

From random vibration theory, the area under the PSD curve is by definition the "mean square" of the response, and the square root of this value is then calculated and is called the "root of the mean square" (RMS), as indicated in Fig. 2.35.

Returning now to the time domain, a RMS of a normally distributed random variable by definition equals one sigma of individual time history points, so the amount of time the response will exist for any specific value can be statistically characterized. If we make this

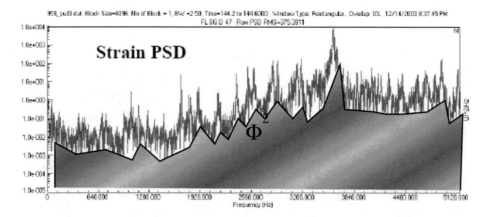

**Fig. 2.35** Mean square of a strain PSD

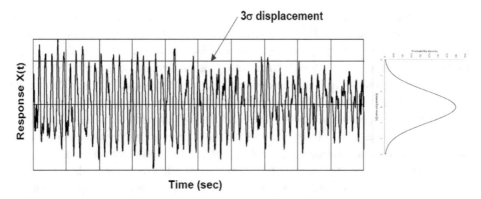

**Fig. 2.36** 3-Sigma value of random response, time history juxtaposed with Gaussian probability density function

assumption of normality, (not perfect, but not bad, either), a commonly used design value for these signals is 3*σ, which covers 99.86% of the responses, as shown in Fig. 2.36.

The procedure laid out here for random analysis has only become tractable in recent years with the advent of structural dynamic analysis capabilities in finite element codes. A simplified technique used extensively in the past and still today for preliminary assessments is to apply a closed form expression developed by J. Miles in 1954 [13]. For SDOF systems excited by white noise, which is defined to be a constant value across all frequencies in a PSD (physically impossible, but frequently approximately true), Miles determined that the response follows a simple closed form equation. This is accurate much of the time because MDOF systems with well-separated modes do behave like SDOF systems, and the white noise assumption only has to exist in the region of the natural frequency. For an acceleration response, the equation is defined as

$$\ddot{x}_{RMS} = \sqrt{\frac{\pi}{2} f_n Q G_b} \tag{2.25}$$

where $f_n$ is the natural frequency in hertz, $Q$ is the quality factor ($1/2\zeta$), and $G_b$ is the magnitude of the PSD of the acceleration white noise.

## 2.7   Modal Testing

Analytical methods have become more and more reliable and high-fidelity, leading to both increased accuracy and precision. Nevertheless, there still is a need for validation of the basic structural dynamic characteristics of natural frequency, mode shape, and damping with a test method, which is called "Modal Test." The values of these quantities obtained through test are used in several ways. Most importantly, the natural frequencies are compared with those predicted by analysis and used to determine if the model is adequate, based on some pre-determined criteria, and if it is not, to guide some future adjustment (generally to the model) to satisfy the criteria. This adjustment, called "modal correlation" is performed either by varying a direct physics-based parameter, such as the Young's Modulus or density, or by using a "non-parametric" technique, where a direct modification of the elements of the mass and stiffness matrices are made. Although the second method seems unjustifiable, there are discussions in the literature suggesting it is more appropriate than physics-based parametric methods [14]. In addition to comparison and correlation, there are new methods in development that use the parameters obtained from modal test to create special finite elements directly. This could have great utility when modeling structures bounded by other structures for which the analyst is not interested, but for which the dynamics will have an important influence through the boundary condition.

These characteristics are heavily dependent on the boundary conditions, though, so it would be logical to assume that modal tests should try to replicate the operational conditions of the component as closely as possible. The problem with this goal is that it is usually impossible to accurately replicate these conditions due to the flexibility of the attached structure, the high speed of rotation (in turbomachinery), the high temperature, the presence of high-density fluids around the structure, or many other conditions.

Therefore, the goal instead in modal testing is to create boundary conditions and excitations that are as easy as possible to replicate in a model. Usually these are free-free boundary conditions, in air, at room temperature, with small, easily controllable, and measurable loads used for excitation of the modes to be measured. Once the match of the model and tested dynamic characteristics meet the criteria, the operational conditions (estimate of actual boundary conditions, rotational speed, operating temperature, fluid environment, and realistic loading) can all be applied analytically. Of course, each of these factors themselves may be predicted and verified by different types of testing to increase the accuracy of the final modal prediction.

The initial technique for modal testing a structure, called "sine sweep testing" and still used for certain applications today, is to attach an electromagnetic shaker to the structure using a thin rod, called a "stinger", which applies a harmonic force (controlled using a force transducer between the shaker and the stinger) that can be slowly "swept" through the range of the natural frequencies of interest. The response of the structure is measured using accelerometers at different points on the structure, and the response can be used to map out a mode shape. This method has in general been replaced by more rigorous methods using a wide-frequency band excitation provided either by a force transducer-instrumented impact hammer or a shaker-stinger. The natural frequencies and mode shapes are determined from the frequency response function, which for modal testing with a force $F$ applied at one location $j$ and response $u$ measured at location $i$, is defined as

$$\overline{F}RF_{ij}(\Omega) = \frac{u_i}{F_j} = \frac{\text{Response at dof } i}{\text{Harmonic excitation at dof } j}$$

With some manipulation, the FRF can be expressed as a function of the modes and the complex frequency response $H$ discussed in this chapter.

$$\overline{FRF}_{ij}(\Omega) = \sum_{m=1}^{M} \frac{\phi_{im}\phi_{jm}}{\lambda_m} \overline{H}_m(\Omega)$$

and further manipulation yields

$$\phi_{jm} = \mathrm{Im}\left(\overline{FRF}_{ij}(\Omega = \omega_m)\right)$$

so to create an entire mode shape, we impact at some point $i$ and collect measurements at the rest of the points (or vice-versa) in all three orthogonal axes (if necessary). An excellent visualization of this calculation is shown in the 3-D graph of the first three modes of a free-free beam, with amplitude of the imaginary part of the FRF tracing out the mode shape in the distance axis, and the different modes identified in the frequency axis (Fig. 2.37, from Zaveri) [15].

## 2.8   Dynamic Boundary Conditions

One of the more difficult aspects of both modeling and modal testing LRE structures is attempting to replicate boundary conditions, especially if the boundary is a structure itself. Frequently the boundary structure is not actually of interest in a dynamic sense (i.e., we aren't interested in its natural frequencies or modes), but its dynamics are an important factor in the dynamics of the structure we are interested in. An excellent example of this

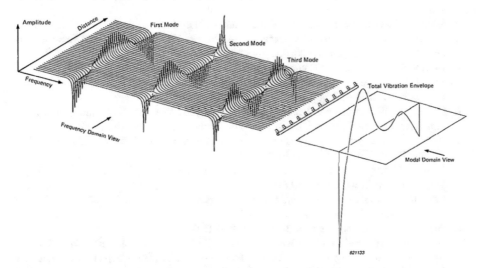

**Fig. 2.37** Schematic of mode in frequency, temporal, and spatial domains

situation is the evaluation of an engine in a hot-fire test stand. The analyst needs to know what the modes of the engine are in that configuration, but he has no interest in the modes of the test stand itself, only in how the test-stand behaves while supporting the engine.

These situations are frequently modelled with springs, in an attempt to simulate the compliance only, and this generally provide reasonable answers. However, the compliance itself is not an accurate boundary condition if modes of the supporting structure are close to modes of the structure of interest. In these cases, the most accurate method for analysis would be to create an entire model of the supporting structure, but this can be classic "overkill," requiring a huge effort. One technique for capturing the boundary condition dynamics without having to resort to modeling the entire support structure is create finite elements based directly on experimental modes [16] or Frequency Response Function [17] taken by modal testing at the support structure interface. For the modal method, mass, spring, and damper scalar elements are connected from ground to NAS-TRAN "spoints" to simulate each measured interface natural frequency. The spoints represent the generalized dofs $\{q\}$ in the standard modal transformation previous discussed, $\{x\} = [\Phi] \{q\}$. The dynamic response of the physical dofs is therefore expressed by generating multi-point-constraint equations directly from each row of the matrix equations defined by the transformation. A more thorough implementation of this concept has been explored by Mayes and Allen in a series of papers discussing combining experimentally based substructures with finite element substructures using the "Transmission Simulation Method" [18, 19].

## References

1. Mars Climate Orbiter Mishap Investigation Board Phase I Report, NASA, November 10, 1999.
2. Rao, SS., Mechanical Vibrations. 3rd Ed., Prentice-Hall, p. 590.
3. Alexander JE, Shock Response Spectrum—A Primer, 27th Conference and Exposition on Structural Dynamics (IMAC) 2009.
4. Blevins RD., Formulas for Natural Frequency and Mode Shape, Krieger Publishing Co.
5. Russel D., Mode Shapes for Multiple Degree-of-Freedom Oscillators.mp4, Penn State University, Available at: https://www.acs.psu.edu/drussell
6. Christensen ER, Structural Dynamic Analysis of Cyclic Symmetric Structures, 46th AIAA Structures, Structural Dynamics, and Materials Conference, Austin, TX, 2005.
7. Brown AM, Comprehensive Structural Dynamic Analysis of the SSME/AT Fuel First-Stage Turbine Blade, NASA/TM-1998-208594.
8. Craig R, Kurdmilla A (2006) Fundamentals of Structural Dynamics, 2nd Ed., John Wiley & Sons, Hoboken, New Jersey, p. 124.
9. Craig R, Kurdmilla A (2006) Fundamentals of Structural Dynamics, 2nd Ed., John Wiley & Sons, Hoboken, New Jersey, Section 16.3-5.
10. Paultre P (2010) Dynamics of Structures. 784. ISTE/Wiley.
11. Nieto, MG, ElSayed MSA, Walch D, Modal Participation Factors and Their Potential Applications in Aerospace: A Review, Progress in Canadian Mechanical Engineering, 30May2018. https://doi.org/10.25071/10315/35254

12. MSC Nastran Dynamic Analysis User's Guide, MSC Software Corporation, p. 578, 2014.
13. Miles JW., On Structural Fatigue Under Random Loading, Journal of the Aeronautical Sciences, pg. 753, November, 1954.
14. Kammer, D.C., Blelloch, P., Sills, J. (2022). SLS Integrated Modal Test Uncertainty Quantification Using the Hybrid Parametric Variation Method. In: Mao, Z. (eds) Model Validation and Uncertainty Quantification, Volume 3. Conference Proceedings of the Society for Experimental Mechanics Series. Springer, Cham. https://doi.org/10.1007/978-3-030-77348-9_15
15. Zaveri K., Modal Analysis of Large Structures, Bruel & Kjaer, 1985, ISBN 87 87355 03.
16. Brown, AM, Ruf J, Calculating Nozzle Side Loads using Acceleration Measurements of Test-Based Models," IMAC-XXV, Orlando FL, Feb 19–22, 2007.
17. Hopkin, RN, Carne TG, "Combining Test-Based and Finite Element-Based Models in Nastran," IMAC-XXII, Dearborn MI 2004.
18. Allen M, Mayes R, Comparison of FRF and Modal Methods for Combining Experimental and Analytical Substructures, IMAC XXV, February 2007.
19. Mayes, RL, Ross MR, Advancements in hybrid dynamic models combining experimental and finite element substructures, Mechanical Systems and Signal Processing, Vol. 31, August 2012, Pages 55–66, https://doi.org/10.1016/j.ymssp.2012.02.010

# Structural Dynamics Applied to Rocket Engine Turbopumps

**3**

## 3.1 Introduction

I find the application of structural dynamics in turbopumps to be one of the most rewarding aspects of the discipline because the essential dynamic characteristics (natural frequency, modes, and resonance) are not only important for further analysis, but critical in of themselves. In particular, direct resonance is the primary failure mode for turbopump flow-path structures. Additional aspects that make the topic fascinating are the close interaction with the turbopump fluid dynamics, both in the gas turbine and the liquid pump, which introduce the criticality of orthogonality between excitation and mode shape, and the cyclic nature of the structures. We'll start by presenting a brief overview of turbopump operation necessary for SD analysis.

## 3.2 Overview of Turbopump Operation

Turbopumps in a rocket engine are a type of turbomachine specifically invented for use in liquid rocket engines. There are a huge number of related machines that use turbines, from turbojets to turbochargers, but turbopumps are generally only used in rocket engines since they are needed to raise the pressure of liquid propellants stored in the fuel tanks from low pressures to the very high pressures required for rocket engine combustion. They consist of two sides, the turbine and the pump. The turbine is driven by hot gas generated by either a gas generator (or pre-burner), which is a mini-combustion chamber, or by hot gas bleeding off from the fuel circuit in more complicated engine cycles (these cycles

---

**Supplementary Information** The online version contains supplementary material available at https://doi.org/10.1007/978-3-031-18207-5_3.

**Fig. 3.1** Impulse and reaction turbine blades

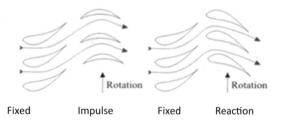

Fixed        Impulse        Fixed        Reaction

require an ignition event to start the whole process, frequently generated by a small solid fuel). The hot gas from this ignition is directed through stationary nozzles or inlet vanes towards turbine blades, which are either installed or fully integrated into a rotating disk. The shape of the blades can either be impulse or reaction, where the impulse blades use a change of enthalpy of the gas to respond (action/reaction, Newton's 3rd Law), and the reaction blades are shaped more like airfoils and use a change in momentum (Newton's 2nd law) to respond (see Fig. 3.1). Turbines can consist of multiple stages to extract the maximum amount of energy from the flow.

The pump consists of a combination of inducers and impellers, and can be classified as either axial or centrifugal, depending on the discharge flow direction. These components add velocity to the flow via either open vanes in the inducer, or shrouded ones, as in the impeller. The inducers are used for the lowest pressure inlets and are designed to raise the pressure enough to eliminate cavitation in downstream stages [1], such as impellers, which are then optimized for the higher inlet pressure flows and produce the extremely high pressures for necessary for combustion. There are a few current designs with combined inducer/impellers. As with the turbine, there are stationary elements in the flow-path, although generally they are only downstream of the impeller in the diffusers, which act to slow the flow down, thereby converting the high velocity to high pressure before it goes into the outlet. Pumps can also have several stages as necessary to obtain the necessary pressure. Cutaway views and cross-sections of the SSME LOX and LH2 turbopumps, excellent examples of advanced engine turbomachinery, are shown in Figs. 3.2 and 3.3 [2].

As rocket engine fuel and oxidizer are cryogenic liquids in the pump sides, special materials are needed that maintain high structural capability while operating at such low temperatures (−423 F for liquid hydrogen). Titanium alloys such as Ti-5AL-2.5Sn are commonly used for liquid hydrogen but are susceptible to ignition in an oxidizer so cannot be used for liquid oxygen pumps, in which Aluminum is used.

## 3.3    High Cycle Fatigue Overview

The main failure mechanism in rocket engine turbomachinery is either low or high cycle fatigue (LCF/HCF). LCF, which is caused by alternating stresses with a cycle count less

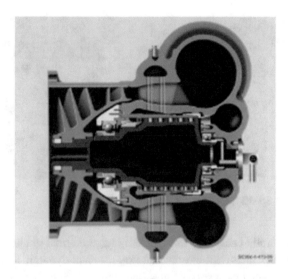

**Fig. 3.2** RS25 low pressure oxidizer turbopump cross section

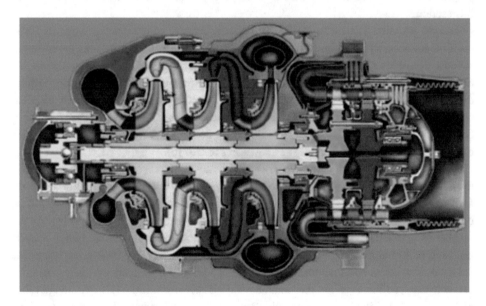

**Fig. 3.3** RS25 high pressure fuel turbopump cross section

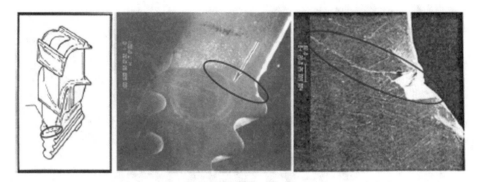

**Fig. 3.4** Cracking of SSME turbine blade, from Min, Harris

than 10,000, is generally caused by unexpectedly large thermal gradients, while dynamic loads cause HCF. It is important to note that non-resonant dynamic loads can cause high alternating stresses, not just resonant ones, and must be evaluated in any development program. But HCF due to resonance is more predominant, and so is discussed briefly in this section.

Fatigue was first studied in 1837, but the term became widely-known when in 1949, the first commercial jet airliner, the de Havilland Comet, had three mid-air catastrophic failures determined later to be fatigue [3]. Wilfred Campbell laid the groundwork for examining HCF in turbomachinery with his seminal work on resonance in steam turbine disk wheels [4]. The basic idea is that a stress much lower than the yield or ultimate strength can cause cracking and failure if it is cyclic in nature. An example of HCF cracking of a turbine blade due to resonance of a bending mode occurred during testing of the Space Shuttle Main Engine, as shown in Fig. 3.4 [5].

Although there has been extensive work on the cause and propagation of fatigue cracks, the determination of a material's fatigue capability remains empirically-based in the "S–N Diagram", which plots the number of cycles to failure $N$ for a material sample for different levels of pure alternating stress $S$, i.e., the mean stress is zero (see Fig. 3.5) [6]. As the number of cycles increases exponentially, the curve starts to flatten out. For some materials, such as Aluminum, it completely flattens at approximately $10^7$ cycles, implying that for alternating stress less than that value there is no limit to how many cycles the material can withstand, so this is called the "endurance limit" giving "infinite life". Although other materials, such as Titanium, do not show that asymptotic behavior, a billion cycles is still generally considered infinite life for design in rocket engine applications.

The next step in a fatigue analysis is to incorporate the effect of mean stress occurring simultaneously with the alternating stress, which is almost always the case outside of the focused testing used to generate S–N curves. A plot of alternating stress versus mean stress is used to examine this capability. Intuitively, it's clear that the y-intercept would be the value obtained from the S–N curve for the required number of cycles (usually the

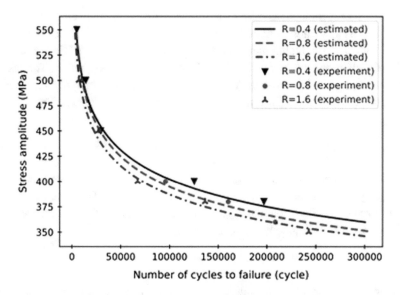

**Fig. 3.5**  SN curve of carbon steel

endurance limit), while the x-intercept is the yield or ultimate strength, depending upon one's failure criteria. It is less clear how these two intercepts should be connected; the "Goodman" criterion uses a straight line, while the "Soderberg" and other criteria use various curves. For either methodology, if the applied combined mean and alternating stress lie below and to the left of the curve/line, the part is acceptable for its dynamic loading conditions, and if it is above and to the right, it is not acceptable and further analysis/testing must be performed (see Fig. 3.6).

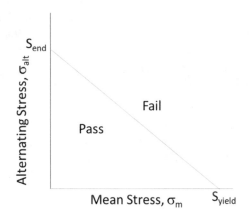

**Fig. 3.6**  Goodman diagram for fatigue

The last major technique necessary for our discussion is defining alternating equivalent stress for use in calculation of finite life damage using Miner's rule. If the component does not meet endurance life requirements, an alternate approach is to use "finite life" analysis, in which the HCF damage is calculated for a given amount of operation. The first step is to calculate $S_{alteq}$, the "Equivalent Alternating Stress", which combines the mean stress and alternating stress into a form that can be used in the S–N diagram. The definition is

$$S_{alteq} = \frac{S_a}{1 - \frac{S_m}{F_{tu}}} \tag{3.1}$$

where $S_a$ is the alternating stress, $S_m$ is the mean stress, and $F_{tu}$ is the ultimate tensile strength (or yield strength, depending upon the requirements). The next step is to determine the cycles to failure for that $S_{alteq}$ directly from the S–N diagram; curve fits for specific materials are frequently available, e.g.

$$N_{fail} = 10^{(-9.2461 \times Log_{10}(S_{alteq})+20.672)}. \tag{3.2}$$

With modern computing capabilities, a continuous application of Miner's Damage Fraction rule can then be applied; the rule calculates the damage $\phi$ at a specific alternating stress value as the ratio of the actual cycles experienced by the component at the value to the number of allowable cycles, i.e., the value on the S–N diagram curve. In the past, cycles were calculated by determining the time within finite speed "bins", e.g. a 50 RPM bin, damage calculated for each bin, and then summed to obtain the total damage.

$$\Phi_i = \frac{N_{accum,i}}{N_{fail,i}}, \quad \Phi_{total} = \sum_{i=1}^{\text{number of bins}} \Phi_i \tag{3.3}$$

With the continuous method, the damage for every cycle is calculated and integrated to obtain the total damage [7]. The first step in this technique is to cull out only the peak values from the stress time history. Each of these values therefore has a cycle count of 1, and along with an automatic lookup of the allowable cycles to failure for that stress value using a closed-form fit of the S–N curve mentioned previously, is then integrated over the entire time history of interest. The speed time history is used, along with an alternating stress time history, and both are then input into the integral

$$\phi(t) = \int_0^t \frac{\Omega(\tau)}{N_{fail}(\tau)} d\tau. \tag{3.4}$$

## 3.4 Characterization of Excitation and Complex Fourier Series

The characterization of the excitation in turbomachinery is as important as characterizing the structure. Since turbomachines are rotating machinery and therefore have a large degree of cyclic symmetry, Fourier Series analysis is essential. We briefly introduced some concepts in Chap. 2, and now will amplify them with a focus on rotating machinery.

The fluid excitations existing in the flow field all have a strong cyclic component, although asymmetries, particularly in inlets and outlets, can disturb this symmetry. Flow fields in turbines and pumps have distortions due to all the component in the flow path; in the turbine, these include inlet guide vanes, nozzles, turbine blades, stators, and exit guide vanes, and in the pump, they include inducer blades, impeller blades, and diffuser vanes. Each of these distortions impart a sinusoidal excitation on the downstream hardware via a change in the fluid potential field, via a wake, sometimes called an acoustic loading, and if the flow is sub-sonic, upstream also via potential disturbance. The fundamental frequency of the excitation is the product of the speed of the engine multiplied by the number of circumferential distortions; this product is called the i'th Engine Order excitation (e.g., 38 Inlet Vanes has a fundamental excitation frequency at 38EO). Computational Fluid Dynamics (CFD) results are the industry standard for providing forcing functions for structural dynamics at this time, and a two dimensional "unpeeled" cross section of a typical flow field for a two-stage turbine is shown in Fig. 3.7. Animation of a similar 2-D flow field for a NASA study is shown in Online Resource 6. A typical complete 3-D mesh domain from CFD is shown in Fig. 3.8.

As an academic example of the excitations in a multistage turbine, take a second stage turbine blade with three primary distortions either upstream or downstream and an overall slight asymmetry due to a radial inlet duct. Each of these distortions will have a different amplitude, so the total pressure field can be expressed as

$$p(t) = b_1 \text{Sin}(t) + b_2 \text{Sin}(2t) + b_3 \text{Sin}(3t) + b_4 \text{Sin}(6t)$$

**Fig. 3.7** CFD results for typical 2-stage turbine section [8]

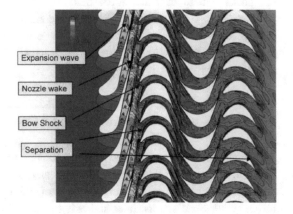

**Fig. 3.8** Turbine CFD mesh
domain

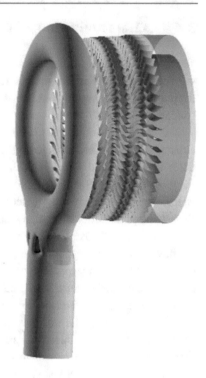

where $b_1 = 0.1$, $b_2 = 0.05$, $b_3 = 1.0$, and $b_4 = 0.25$. The total pressure field superimposed on its components is shown in Fig. 3.9.

Rather than try to assess the structural response of the blade to the total field all at once, we instead calculate the response to each harmonic component.

A single stage turbine perhaps can be considered the "simplest" flow field, with just one distortion propagating upstream and one downstream. However, even these distortions cannot be represented by single harmonic waveforms. As presented in Sect. 2.3.5, the pressure field or nodal force field will instead best be represented by a sum of Fourier terms. The magnitude of these components can be shown in the frequency domain (Fig. 3.10). We will be returning to this topic with a focus on the complex Fourier representation with spatial phase later in this chapter during the discussion of excitation of nodal diameter modes.

## 3.5   Finite Element Modeling of Components

We stepped through an example of discretizing a simple structure into finite elements in Chap. 2. Some discussion on more complicated structures such as turbomachine components is now necessary. In general, models of turbomachine flow-path structures, such as turbine blades and stators, inlet and exit guide vanes or nozzles, inducers, and impellers

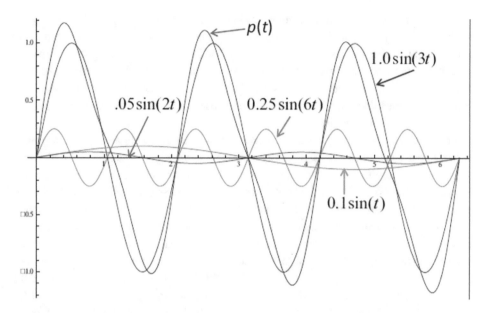

**Fig. 3.9**  Total pressure field as sum of components

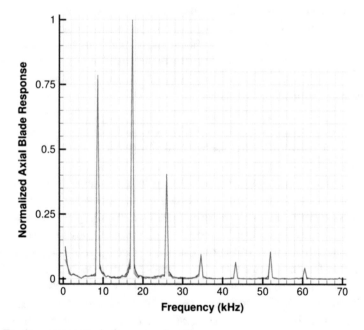

**Fig. 3.10**  Fourier components in frequency domain

require 3-D models using solid elements. However, the axiom "if it looks like a plate, model it with plates" still holds for very thin structures, especially if stress results are being sought. If there is an interface of the shell with a region requiring solid elements, special care must be taken when joining the parts, as there are a different number of basic degrees of freedom per node defined for shells (5) versus solids (3). Most FE codes have special techniques that must be adhered to for a successful connection, e.g. in Ansys, the "node merge" option must be used.

There have been other common concerns which in recent years are slowly becoming of less importance. First, it was generally preferable to attempt to use hexagonal elements for solids instead tetrahedral (tet) elements, as shown in the turbine blade model for the analysis introduced in Sect. 2.5.3, shown in Fig. 3.11 [9], but previous drawbacks in tet meshing have largely been overcome, and automatic meshing of solid geometries is generally much easier using tet elements. In addition, for turbine and pump housings, an axisymmetric 2-D mesh was created to save computational resources rather than creating a full 3-D mesh, using special axisymmetric elements. As CAD geometries are now directly imported into FE packages and meshed, this is becoming less important. Finally, for quarter-symmetric, half-symmetric, or fully axisymmetric geometries, a method for obtaining a good visualization of the modes while still obtaining significant computational savings was using a quarter-modeling technique; it can be proven that all the modes for these geometries can be obtained by 3 models composed of symmetric-symmetric boundary conditions, symmetric-antisymmetric, and antisymmetric-antisymmetric at the 0 and 90° faces. These methods are all worth being aware of for various reasons including a very expensive computational cost such as in a probabilistic or sensitivity analysis, where thousands of simulations may be required. In general, cyclic symmetry is used most of the time, though, as already mentioned.

## 3.6    Modal Analysis and Campbell Diagrams

A modal analysis is almost always the first step in a structural dynamic analysis of a turbopump component. Not only is it necessary for the modal superposition method of response analysis, but in turbomachinery, in particular, knowledge of the natural frequencies and mode shapes themselves is critical as resonance is possible. The steps for an academic modal analysis were laid out in Chap. 2 to give the reader a grasp of the physical concepts, but for finite element models of realistic structures with many thousands of DOF's, numerical methods are used instead. The use of these methods is generally transparent to the analyst, with the technique automatically chosen by the commercial finite element code, and so is not included in this text, but details can be found on any textbook on numerical methods for large eigenvalue problems, such as the text by Saad [10].

One option that is still available for the analyst is whether to use lumped-mass or consistent mass representations in the solution. Lumped mass is self-explanatory and is

**Fig. 3.11**  Hexagonal finite
element model of space shuttle
main engine high pressure fuel
turbopump 1st stage turbine
blade

the technique presented in Chap. 2. A more accurate method, especially for stress or
strain, is consistent mass, which distributes the mass using the shape function in the finite
element formulation. For a reasonably dense mesh, the lumped-mass approximation is
adequate for natural frequencies and mode shapes, and it runs faster, so that trade should
be kept in mind.

It is critical in modal analysis to specify the actual rotation speed, mean pressure field,
and operation time-consistent temperature (since the temperatures will change rapidly
upon startup and then adjust during the various phases of operation) in the static analysis
preceding the modal and frequency response steps. The first two of these parameters will
impose an initial stress field which will non-trivially alter the natural frequency. This is
somewhat counter to our understanding of basic structural dynamic theory which says
that the modal analysis is a solution of the homogenous equation in which the forcing
function is defined to be zero. For these loads, though, the stress field changes the stiffness matrix itself, so the modes will be different. The time-consistent temperature is also
critical as Young's Modulus is highly dependent on the temperature, resulting in altered
natural frequencies as well. Implementing the correct Young's Modulus as a function of
temperature on an element-by-element basis will require a table look-up capability within
the finite element code.

### 3.6.1   Simple Campbell Diagrams

The most prevalent tool in use for modal analysis of turbomachinery is the Campbell Diagram, which was presented at the ASME annual meeting in 1924 by Wilfred Campbell [11], who sadly died just a few months afterwards. Although in its simplest form it could be considered just a graphical method for showing coincidence of an excitation frequency in a rotating machine with a structural natural frequency, i.e., resonance, it actually is much more complicated if considered in more detail.

Let's start by examining the simple case, the Campbell Diagram for the turbine blade discussed in Sect. 3.5 as shown in Fig. 3.12. The abscissa has units of revolutions per minute (RPM), and the ordinate is frequency in hertz (cycles per second). The diagonal lines identify excitations as determined by the configuration, which are the frequencies of multiples of engine speed corresponding to any flow distortion in the flow-path, e.g., the number of upstream stators adjacent to this turbine stage. The slope of these lines is shallow because of the factor of 60 between Hertz and RPM, i.e., if both axes were in the same scale an engine order of 26EO would have a slope of 26, but for a Campbell Diagram it has a slope of 26/60. The excitations include (but are not limited to) multiples of the primary distortion number depending on the strength of the Fourier harmonic, as discussed previously. The vertical dashed lines are the bounds of the operating speeds of the turbopump. In this case, the blades are mounted on a very thick forging instead of a typically thin disk and therefore are independent of each other, so the horizontal lines are the cantilevered structural natural frequencies of the blade. Resonance occurs in this case when a diagonal line crosses a horizontal line within the bounds of the operating range (between the vertical lines). This condition represents an oscillatory excitation acting on the blade as it moves into and out of circumferential flow distortions at its natural frequency. One advantage of a Campbell Diagram is that it can incorporate changes in the natural frequencies due to operating speeds, which is a small but non-trivial increase due to a nonlinear effect called centrifugal stiffening. If the operating speed is tied to a specific time in the engine operating profile, it can also imply a change in temperature of the component, which can have a non-trivial effect as well. Both of these adjustments can be included by running a static analysis first that will incorporate these effects into the modal analysis. The final result will be curves instead of a horizontal lines for the natural frequencies. We will discuss other conditions necessary for more general blade/disk resonance and resultingly more complex Campbell diagrams in a later section.

### 3.6.2   Margins for Resonance

The definition of what exactly constitutes resonance can be very controversial in an engine development program. If there were no uncertainty or variability, the question could be answered easily by assigning some critical percentage of the peak response as calculated

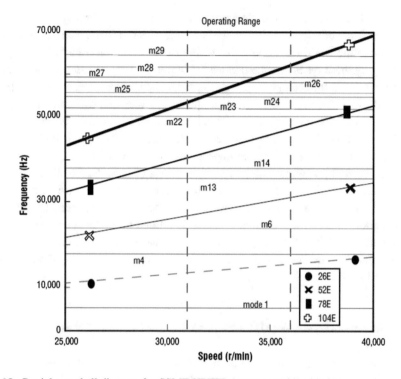

**Fig. 3.12** Partial campbell diagram for SSME HPFTP 1st stage turbine blades

using the SDOF frequency response Eq. 2.10 (realizing that this is an approximation of the MDOF response) and calculating the frequency band for this level. A common bandwidth to use would be that corresponding to the half-power values discussed in Sect. 2.3.3, where a typical damping value would be assumed, e.g., for $\zeta = 0.002$, the half-power bandwidth defined by the range $f_1$ to $f_2$ is

$$f_2 - f_1 = 2\zeta f_i = 2(0.002)(20,000\,\text{Hz}) = 80\,\text{Hz},$$

so the percentage variation would be $80/20000 = 0.004$, or $\pm 0.4\%$.

In real systems, though, there is always variability and uncertainty. A number of excellent texts have been written examining the topic of uncertainty quantification, particularly the one by Coleman and Steele [12], and gaining a working knowledge of this material, along with background statistics, is critical for SD of LRE's. Establishing definitions of these terms is vital; variability is the mathematical definition of a non-deterministic process, while uncertainty is described by a certain type of variability according to whether it is "aleatory" or "epistemic." Aleatory uncertainties are due to natural, generally irreducible variabilities in material properties, flow conditions, or other physical phenomena. Epistemic uncertainties are due to a lack of knowledge or inability to model the physics accurately. Of course, there's the old axiom, "Everyone believe the test results except for the tester, and nobody believes the analysis results except for the analyst," but even the best analyst in the world must admit that there are modeling uncertainties.

Quantifying these types of uncertainties are important for generation of and adherence to a margin criteria for avoidance of resonance, i.e., how much percentage does the excitation need to be away from the natural frequency. A very quick green flag is generally considered to be 20% for a totally analytical prediction. Modal test and other additional sources of knowledge (e.g. measured temperature conditions, which will significantly affect natural frequencies due to the dependence of Young's Modulus on temperature) can then be used to justify moving to a margin criteria of between 5 and 10%, depending on the confidence in the modeling, test, and the severity of consequences if the part were to fail—for instance, a larger margin may be used if human life depended on the outcome rather than downtime for a generator. It is also important to establish a criteria that is not so conservative that it will be impossible to meet, and therefore ignored. For any resonance margin, the outcome of violation will require some sort of mitigation, which will be discussed in Sect. 3.6.9.

### 3.6.3    Mode Shapes, Modal Animations and Modal Stress

Since failure at resonance is a major failure scenario, detailed knowledge of modal response, including modal animation (video of a modal deformation vector over a cycle) and modal stress, are extremely valuable in turbomachinery analysis, perhaps more so than most other applications in structural dynamics of space structures. All commercial FE codes can generate modal animations very easily. Although a contour plot of a mode shape can give the same information as an animation, it just isn't the same—it'd be like watching a movie as a collection of still images. An animation will be able to show the relative phase of the different locations in the structure far better than any plot, which is critical for identifying how excitations will interact with the mode shape. Modal animations are also critical for placement of external dampers (to be discussed in Sect. 3.7), to see clearly the locations of maximum displacement; good examples for NASA turbine blades are shown in Online Resources 7 and 8.

Modal stress (or strain if appropriate) is also a valuable quantity not frequently used elsewhere in structural dynamics of space structures. It is simply the stress field on a structure if the normalized mode shape is enforced. As with the modal deformations, the actual values of the stresses are not meaningful, only the relative values. These relative values are critical for HCF determination, as they determine where the highest stresses would be if the response is dominated by a single mode. A modal stress plot of the suspect mode in the failure analysis of the SSME turbine blade clearly showed the peak stresses occurring at the location of crack initiation, helping to verify that the mode was a contributor to the failure (Fig. 3.13).

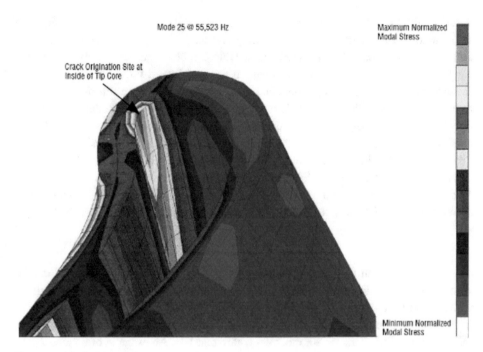

**Fig. 3.13**   Mode 25 modal stress plot showing peak stress at crack initiation location

## 3.6.4   Implications of Cyclic Symmetry in Turbopumps

### 3.6.4.1   Interference Diagrams

We discussed the basic concepts of cyclic symmetry modeling in Sect. 2.6. The implications of cyclic symmetry in turbomachinery lead to some of the most complicated and unique aspects of structural dynamics of LRE's. The basis of these factors is the property of orthogonality, mentioned in Eq. (2.18) with the mathematical integral shown below:

$$\int_0^{2\pi} \sin(n\theta)\sin(m\theta)d\theta = \begin{Bmatrix} \pi \text{ for n = m} \\ 0 \text{ for n} \neq \text{m} \end{Bmatrix} \tag{3.5}$$

Equation 3.5 implies that in order for any appreciable resonant response, not only must the excitation frequency be close to the natural frequency but that the circumferential wave number of the excitation shape must equal the wave number of the mode shape. In the simplest scenario, a flow disturbance caused by 26 stator vanes generates an excitation shape with 26 full waves around the periphery, and with Fourier components at multiples of 26. These disturbances can only excite modes with an equal number of waves, or nodal diameters, as defined in Sect. 2.6 (although this is not really entirely true due to the "Tyler-Sofrin" complication, which will be discussed shortly). The "simple" Campbell Diagram described the Sect. 3.6.1 no longer is valid in this general case, but there are a number of ways to correct the diagram. The most straightforward is to create a Campbell for each and every excitation, and only include modes that can be excited by that excitation, e.g., for the 26-stator case, the Campbell diagram for an adjacent turbine stage would only include modes that fall into the 26 nodal diameter family. There are still plenty of modes to worry about, as each of the "blade-only" modes exist for each nodal diameter. "Blade-only" modes are loosely defined as modes where the turbine blades are in a cantilevered configuration, or where the disk portion is considered fixed as was the case for the huge disk in the SSME alternate turbopump as previously described.

A "nodal-diameter" plot of the modes is critical for this analysis. In this plot of the mode shapes of a turbopump cyclically symmetric component, the abscissa is the integer number of nodal diameters, and the ordinate is the frequency. The maximum ND value is

$$ND = N/2, \text{ or } (N - 1)/2 \tag{3.6}$$

A line is drawn in the graph connecting the modes with the same sequential number within each nodal diameter, and these are sometimes called the same "modal family", e.g., the first bending family, where the turbine blade simply bends about its most flexible axis. A specific mode is referred to by its nodal diameter and sequence number, e.g., mode 20,6 is nodal diameter 20, mode #6 within that ND. An example of a mode 2,1 for simple integrally bladed-disk is shown in Fig. 3.14, and the nodal diameter plot is shown in Fig. 3.15.

The nodal diameter diagram only contains natural frequency data, not any excitation data, but it is valuable in a number of ways by itself. First, the value at the highest nodal diameter is generally considered to be the "blade-alone" frequency, in which almost all the strain energy of the mode is concentrated in the airfoils rather than the disk or shroud. Second, locations in the diagram where the lines of different families come together and then veer apart illustrate a frequently-occurring phenomena called "eigenvalue veering." For bladed-disks, these regions have been found to generate large values of mistuning amplification, which will be discussed in Sect. 3.6.7 [13].

When creating a cyclic symmetric finite element model, the cyclic boundaries must be clearly identified and labelled as side "a" and "b", so that the program knows to connect side "b" of an adjacent sector (created numerically inside the code) with side "a" of the modeled sector, and similarly that side "b" of the modeled sector connects with side "a" of the other adjacent sector. As mentioned previously, these boundaries do not have to be planar, and should either be created to give the best view of a single sector (although most codes now have the capability to create other sectors to enhance visualization), or just chosen according to the easiest x-section to isolate.

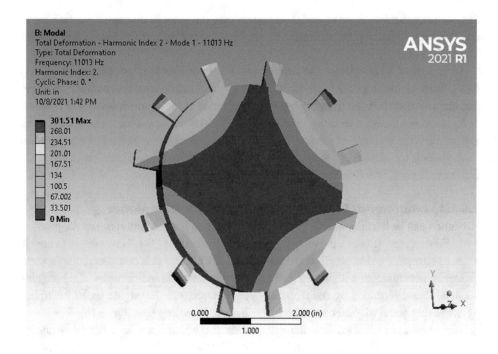

**Fig. 3.14** Simple blisk mode 2, 1

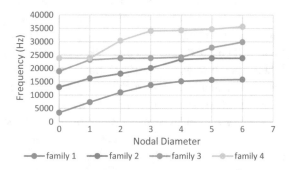

**Fig. 3.15** Simple blisk nodal diameter diagram

### 3.6.4.2 Blade-Dominated Nodal Diameter Modes Versus Disk-Dominated Nodal Diameter Modes

Two different types of nodal diameter mode shapes exist in turbomachinery, and proper classification of the type is critical for correct analysis. For turbine-side bladed-disks, also called turbine wheels, the higher nodal diameter modes can only be seen in the response of the blades, not in the disk portion of the component. Therefore, a Nyquist-type limitation that the maximum number of nodal diameters that can exist is N/2 or (N − 1)/2 for an odd number of blades is because not enough "data points" exist to define a wave number higher than that.

On the other hand, pump-side components have portions that are almost completely disks, such as the impeller back-face. There is no limitation on the number of waves that can exist, even if the impeller has a finite number (usually much lower than bladed-disk) of cyclic-sectors due to its internal vanes. Therefore, an excitation of spatial order greater than the number of sectors can still excite the higher-order wave number. The methodology for identifying the possible excitations is by a slight alteration of the nodal diameter diagram into the "interference" diagram.

### 3.6.4.3  Interference Diagram and Campbell for Disk Modes

Excitation can be introduced into this diagram, changing it to an "interference diagram" or a "SAFE diagram" (Singh's Advanced Frequency Evaluation) [14]. The abscissa is now in terms of harmonic index, which referring to Sect. 2.4.7, Table 2.2, contains multiple nodal diameters according to a stair-step pattern. The excitation is indicated by a zig-zagging diagonal line determined by the frequency corresponding to the engine speed multiplied by increasing nodal diameter for each of the harmonic indices (see Fig. 3.16). Resonance occurs when the excitation line lies on a mode for which there is a spatial distortion (or Fourier harmonic) for that nodal diameter. The speed will be either defined over a range and there will be two excitation lines, or there can be a variability about a single speed, and resonance will occur when the line crosses the dots signifying the modes. This representation recognizes the orthogonality implication that the gross character of the excitation shape along the circumferential direction must share a wave number with the mode shape, as signified by the nodal diameter. For example, if there are 5 flow distortions in this case, then there will be excitations in a 5ND shape as well as harmonics of 5, i.e., 10, 15,…. The 10ND excitation would excite 10ND modes, which would be in the second Harmonic Index. The convention is to draw a short vertical line at the excitation within the operating range, so this plot shows there is not a resonance for this specific excitation. It is helpful also to identify the nodal diameter value of the excitation along the right ordinate, especially for large nodal diameter values. In the "Safe" diagram commonly used, the abscissa is identified as "nodal diameters" instead of "harmonic index". This is accurate when considering excitation of bladed disks (further excitation mechanisms to be discussed in the next section), but for disk-dominated modes, this somewhat confuses the Harmonic Index with the Nodal Diameter.

In this example, the plot shows crossings at 2, 4, and 6 nodal diameters (which would be a very poor design!). This is because "best practice" dictates that 1–6 engine order always be applied on a component with a possible disk mode. There will always be an excitation at 2EO and its harmonics 4EO and 6EO due to unavoidable slight rotor imbalance, but we also consider the odd numbered excitations. It has been demonstrated repeatedly (the cases are proprietary) that these very fundamental disk modes can

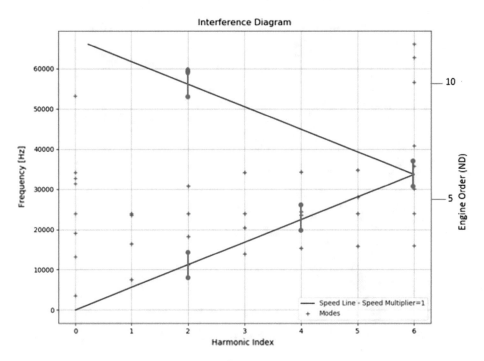

**Fig. 3.16**   Interference diagram for simple blisk

respond if this interference occurs, even if there is no known source of excitation (private discussions with Dr. Robert Kielb, April 2018).

One significant benefit of the interference diagram/cyclic symmetry technique is that by looking at only the modes for specific harmonic indices, a filter is applied to the modal extraction to yield only nodal diameter modes in that HI, in the order indicated by the HI table. This provides an automatic ND characterization of the mode rather than relying on examining the mode shape itself, which can be tedious and error-prone.

On a final note, yet another method for showing disk-mode resonance in a Campbell Diagram is to superimpose a marker on the horizontal line of the natural frequency where an excitation order crosses a mode with the same nodal diameter. It is only at this crossing where true resonance exists, not when that particular excitation order diagonal crosses the horizontal lines elsewhere. Since this is a crossing of the excitation, the modal natural frequency, and the mode shape, this is called a "triple crossover;" and Campbell called it a "disk critical speed." This method is still very commonly used for impellers. An example is shown in Fig. 3.17 for the NASA LPS breadboard engine, where the natural frequencies of the modes are represented by bands accounting for uncertainty, and the appropriate "double", but not "triple" crossovers are indicated by boxes. As can be seen, there is a huge margin for both the 1ND and 2ND modes. An additional complication

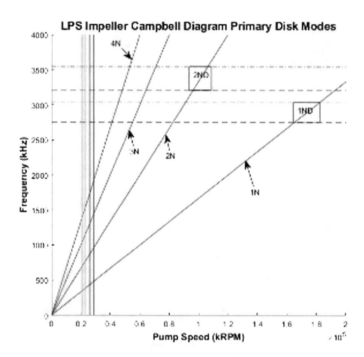

**Fig. 3.17** LPS impeller campbell of primary disk modes

for impellers is that there are both internal vanes that extend the full spiral length and vanes that start halfway down the spiral and only extend half the length (Fig. 3.18). Excitation of and by these half-vanes should only be considered for downstream, not upstream distortions, since the wake disturbance from upstream will have attenuated by the time it reaches the half-vanes.

### 3.6.4.4   Traveling Waves Versus Standing Waves

At first glance, the excitation mechanism previously described in this section seems straightforward. Upon closer examination, though, we see that the excitation pressure fields propagating either upstream or downstream are not standing waves but are rather traveling in the reference frame of the component of interest. In other words, rather than the maximum amplitude of a wave staying stationary on a certain location, it rotates. For the turbine, the direction of the excitation wave is the opposite direction of rotation. Since the mode shapes derived in a modal analysis appear to be stationary, this would seem to be a problem. In fact, as mentioned previously, all the nodal diameter modes in cyclic symmetric structures actually exist in pairs. The circumferential spatial solutions for these can be written as

$$\Phi_1 = \sin m\theta \qquad (3.7)$$

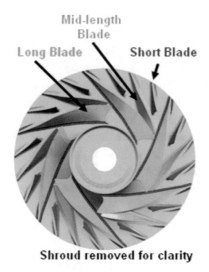

**Fig. 3.18**  Typical impeller with full and half vanes shown (Shroud not shown)

$$\Phi_2 = \cos m\theta$$

for which the total modal solution $u$ for both modes vibrating simultaneously at frequency $\omega_m$ is

$$u = u_1 + u_2 = A\sin(m\theta)\sin(\omega_m t) + A\cos(m\theta)\cos(\omega_m t).$$

This is equivalent to

$$u = A\cos(m\theta - \omega_m t)$$

which is a wave traveling at speed

$$\dot{\theta} = \frac{\omega_m}{m}.$$

Since the signs of Eq. 3.7 are arbitrary, the sign of $\omega_m t$ is also, so there is also a traveling mode going in the opposite direction, i.e., backward traveling and forward traveling waves (relative to the direction of rotation). The direction of the excitation will determine which of these modes is excited. For turbines, therefore, the backward traveling mode is excited, while for stators, the forward traveling wave is excited.

## 3.6.5  Tyler-Sofrin Modes

A very interesting feature of the specific mechanism of excitation in the flow path components of turbomachinery are the excitation of modes in nodal diameter families not equal to the number or harmonics of excitation flow-distortions. These types of interactions were first identified in a seminal study by Tyler and Sofrin [15], and so are sometimes referred to as "Tyler-Sofrin modes." These modes are tied to the acoustic propagation properties of the turbine, so are also sometimes called "acoustic modes", but the important SD characteristics are a simple function of the number of excitation distortions (e.g., the number of stators), and the number of receiving components (e.g., the number of blades). For this case, the number of excitable nodal diameters $m$ for $B$ blades excited by $V$ stators is determined by

$$m = nB + kV \tag{3.8}$$

where $n$ is the integer Fourier coefficient of the excitation, going up to a value that CFD shows has a non-trivial amplitude, and $k = ..., -1, 0, 1, ....$. The physical cause for this unexpected interaction is spatial aliasing and modulation of that aliasing; since only a finite number of locations actually feel the forcing function, unlike a continuous disk, these locations can only sample the excitation periodically. This sampling causes the excitation nodal diameters to be converted into other traveling nodal diameters. The direction of the traveling wave is given by the sign of the $m$, where positive are "forward traveling waves" and negative are "backward traveling waves". An example of 69 blades being excited by 74 vanes is shown in Fig. 3.19 and Online Resource 9, an animation in which the pressure wave is progressing from left to right, causing the aliased 5ND wave ($69 - 74 = -5$) to travel in the reverse direction. This wave number can be seen in the figure where the excitation wave is the green continuous curve while the theta location and pressure applied onto the blades are signified by circles. Connecting the circles clearly subtends a 5ND diameter waveform. As described in the previous section, the corresponding traveling wave mode will respond, but it is critical to note that the temporal frequency of the excitation is equal to the number of excitation distortions.

It is certainly not intuitive why there is an integer multiple of the receiving component. A thorough examination is presented by Figaschewsky [16] showing that the fluid field between the components is the critical region, and the Fourier components of both up and downstream disturbances will affect this region equally. Kushner [17] suggests that integer values of only 1 and 2 should be considered, but the author's experience has shown that substantial Fourier components of the exciting component exist up to the 7th harmonic for sharp edged nozzles. Experimental evidence of these modes for the first harmonics ($k = -1$ and 1) was first presented by MacBain in 1984 [18], but a thorough examination of how many receiving multiples are necessary has not been located in the literature.

A typical method for easily determining the Tyler-Sofrin modes is to create a table like the one in Table 3.1 for the turbine in the example given above. This table only contains

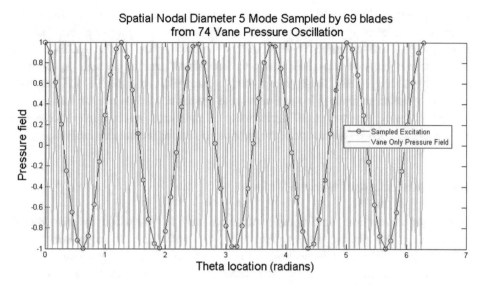

**Fig. 3.19**   Tyler-Sofrin 5ND family generated by 69 Blades excited by 74 Vanes

**Table 3.1**   Direct and Tyler-Sofrin nodal diameter chart for 74 vanes → 69 blades

| Upstream vanes | Direct excitation | 69 blades | 2 * 69 = 138 | 3 * 69 = 207 |
|---|---|---|---|---|
| 74 | ~~74~~ | −5 | 64 | ~~133~~ |
| 148 | ~~148~~ | ~~−79~~ | −10 | ~~59~~ |
| 222 | ~~222~~ | ~~−153~~ | ~~−84~~ | −15 |

the difference calculation from Eq. 3.8. since the sum calculation will almost always result in modes higher than the spatial ND Nyquist cutoff. In this case, that Nyquist maximum number of definable nodal diameters is (69−1)/2 = 34, and the entries above that number are crossed out. It is recommended that the chart also include the directly excited (by wakes or potential change) nodal diameters so that the chart completely accounts for all possibilities.

### 3.6.6   Multi-stage and Asymmetry Effects

Figaschewsky also considers the effect of multiple stages on nodal diameters that are potentially excitable. Theory and experiment have shown that one of these ND's is equal to the difference between the number of blades on different stages, and another equal to the difference between the number of stationary stators or vanes. This topic is an active area of research [19]. Hegde et al. show that the nodal diameter field can be much more populated that the Tyler-Sofrin charts suggest when full 3-D flow fields of multi-stage turbines are modeled, and that up and down-stream effects can interfere with each other in either a constructive or destructive manner. Brown and Schmauch showed that asymmetric inlets also can generate non-trivial energy in "sidebands" near to the main Tyler-Sofrin mode [20]. Although these findings imply that nothing short of a full 3-D CFD analysis is required, the research does indicate that some special conditions need to exist for the response at these secondary nodal diameters to be significant, so the methods described in the previous sections are adequate for most cases and for preliminary analysis.

### 3.6.7  Mistuning of Bladed-Disks

There is yet another complicating factor which we must now introduce, the phenomenon of mistuning. This phenomena is critical as it alters both the basic modal structure of cyclically-symmetric structures as well as the resonant response. Mistuning comes about if the cyclic-symmetry is altered, usually unintentionally, due to manufacturing or material deviation. It is particularly notable in turbine bladed-disks. The phenomenon has been a favorite topic for Ph.D. dissertations for the last 50 years, but reasonably accurate techniques for predicting the response of a specific mistuned bladed-disk, which is the "deterministic problem", have only existed since the creation in the early 2000's of the Subset of Nominal Modes Method by Yang and Griffin [21]. On the other hand, Griffin and Kielb's Fundamental Mistuning Model, first presented in the 80's by Griffin, Kielb, and others [22], generates a reasonable approximation of the response for a population of bladed-disks (non-deterministic mistuning), which is the case for general design particularly if frequency testing of every disk is not planned.

Essentially, when there is a slight deviation between each sector (generally turbine blades), there is a "localization" of the modal energy to a small subset of the blades or even a single "rogue" blade. The resulting mode, which is quite warped, appears almost nothing like the original "tuned" mode (when all the blades/sectors are assumed to be identical). As a result of the localization, the resonant response of the "rogue" blade is significantly higher than the response of the tuned system, up to 200% larger! The value of this "mistuning amplification factor" (MAF) varies widely depending on the amount and location of the deviation and the degree of coupling between the blades through the disk, as reflected in the modes. Reasonable estimates of modes that could have problematic MAF values can be obtained by examining the interference diagram; locations where two or more modal families approach and then veer away from each other (a phenomena called "eigenvalue veering", which is also present in other structural dynamic analyses) are generally the highly-responding cases [13]. Of course, it should always be kept in mind that the mode has to have some reasonable amount of response to begin with, even in the tuned system, in order for the mistuned system to be a problem. For example, in Fig. 3.12, this veering exists at the approach of family 4 to family 3 at nodal diameter 1. Eigenvalue

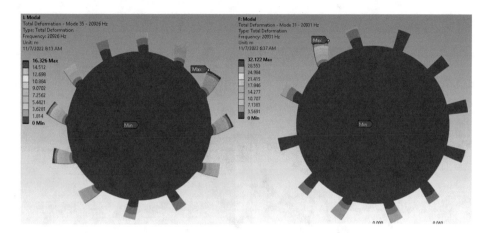

**Fig. 3.20** **a** Nominal (Tuned) Mode 2, 1 of Academic Blisk; **b** Mistuned Mode

veering is clearly seen in Fig. 3.15 for both ND 1 and ND4 at around 23,000 Hz. The effect of the mistuning on the mode 2, 1 shape itself is shown in Fig. 3.20.

Modeling of mistuning is usually accomplished by applying an assumed statistical distribution on the blade stiffness via the Young's modulus assignment on each blade. This technique, called "stiffness mistuning," assumes that the mode shapes do not vary, which is not always true. Methods which assign variation to the geometry itself, called "geometric mistuning," do exist but are significantly more complicated, as one would expect [23].

The deterministic resonant response to a unit pressure loading on the mistuned simple model from Fig. 3.13 for two bladed-disk samples (each with different assumed variations of Young's Modulus) is shown in Fig. 3.21. The analysis was performed in ANSYS© using the Component Mode Mistuning Method [24]. Details of this methodology are not discussed here. The effect on the forced response, seen by examining the amplitude of the peaks for each curve, yields a MAF of about 1.55, i.e., a 55% increase in the predicted forced response compared to the nominal "tuned" system. We will discuss forced response in more detail in Sect. 3.8.2.

If measurements of the mistuning variation, which generally consist of natural frequency testing of each bladed-disk, is not made part of the standard verification procedure, then a non-deterministic mistuning analysis must be performed that will yield statistics of the maximum response. A software code such as Duke University's MISER, which uses the Fundamental Mistuning Model Method, can perform this analysis [25]. As with CMM, the details of this method are found in the reference and other literature. A 2000 sample for this case yielded a 99.865 percentile exceedance MAF value (equivalent to 3 sigma for a normal distribution) of 1.578. These 3 sigma results in this case are similar to the CMM deterministic values, indicating a significant underprediction in general; this is

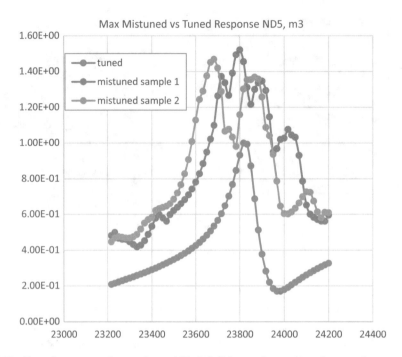

**Fig. 3.21**   Forced response of two mistuned bladed-disk samples compared to tuned response

not unexpected as this version of MISER was only considered accurate in isolated regions of the interference diagram. Improved versions of the code are under development which will significantly increase the applicability of the code to other regions of the diagram.

### 3.6.8   Modal Analysis of Pump-Side Components

The effect of operation in liquids in the pump-side adds tremendous complexity to the generation of the modes and natural frequencies of pump-side flow-path components. NASA/MSFC has completed and published some of the results of a multi-year test/analysis program on the RS-25 inducer (see Fig. 3.22) to examine each of the effects in detail [26]. The first major effect is the increase in natural frequency due to higher Young's Moduli at cryogenic temperatures. Testing at these temperatures is not extensive, but some data has been published, as discussed in the referenced paper.

The reduction in natural frequency in liquids due to the effective added mass of the liquid moving along with the structure is much more significant. This effect has been studied analytically, numerically, and experimentally beginning with Lindholm in 1965 and continuing until the present [27]. Although Lindholm's experiments were in water, as is the case for almost every other experimental study, he did generate an equation relating

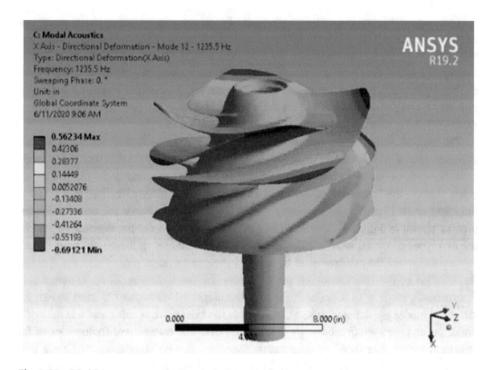

**Fig. 3.22** RS-25 low pressure fuel pump inducer mode shape

the density of the liquid to the natural frequencies, where the same equation is applicable for all cantilever beam-type modes, as derived below

$$A_{m1} = \tfrac{\pi}{4}\rho_f a b^2$$

where $A_{m1}$ is the added mass for beam-type modes, $\rho_f$ is fluid mass density, $a$ is the beam length, and $b$ is beam width. This added mass can then be used to calculate the ratio of the natural frequency of the beam immersed in fluid $\omega_f$ to the natural frequency of the beam in vacuum $\omega_v$ (or a low-density gas) using

$$\frac{\omega_f}{\omega_v} = \frac{1}{\sqrt{1 + \frac{A_{m1}}{m_b}}}.$$

Although the analytical methods provide a good answer for very simple cases, they become increasing less accurate for more complicated structures. The MSFC work has proven that any reasonable estimate of the actual dynamic characteristics can only be determined using a numerical code with acoustic finite-element capability (a computational fluid dynamics run is not necessary).

Two other effects on the dynamic characteristics of rotating structures in the pump side are also important. The first is the effect of tight tip clearance between inducer blades and the pump housing. As can be intuitively imagined, a tight clearance prevents easy flow of the liquid around the tip as the structure vibrates, so more of the liquid is entrained on it, thereby increasing the effective mass. Xiu, Davis, and Romeo performed an analytical, numerical, and experimental study on this effect which was also confirmed by the MSFC work [28]. They also looked at the effect of clearance on damping.

The second major effect is the interaction of fluid acoustic and structural modes, explored extensively by Davis' dissertation [29]. In low-density liquids like LH2, purely acoustic modes (i.e., the structure is assumed to be rigid) at some nodal diameters will come close to the frequency of the structural modes (with added effective fluid mass) at the same nodal diameter. A frequency versus circumferential wave (or nodal diameter) plot, as shown in Fig. 3.23 for the simple cylinder examined for the dissertation, shows the two types of modal families. Davis discovered that neither of these pure modes actually exist and that they are replaced by acousto-structural system modes above and below the location of the intersection of the two lines. In this figure, A(X, n, Y) refers to a pure acoustic mode with X nodal circles, n circumferential waves, and Y axial waves; the structural modes are denoted S(X, n), where X is the axial wave number and n is the nodal diameter; and the system modes are the curved lines. The true natural frequencies of the structural modes that could be of concern, therefore, can be significantly different than those modes even with the effective added mass. To complicate matters further, the bifurcation of the system modes means additional modes must be examined. MSFC's work verified this phenomena for inducers, and further found that the bifurcation can lead to up to 4 separate modes at the same nodal diameter whose structural component of the mode shape all appear to be identical, and therefore all could be resonated by the same excitation shape. This phenomena shows the absolute necessity of performing an acoustic-structural finite element analysis and being aware of bifurcated modes at frequencies different from just using the fluid effective added-mass.

Finally, another of the major challenges of performing and characterizing the modes of inducers is that the blades display both a "blade-like" and "disk-like" character. This can lead to confusion about whether blades should be characterized by their nodal diameter, with only similar excitation shapes able to generate a resonant response, as discussed in Sect. 3.6.4.1. During the initial phases of the MSFC inducer project, this concept was believed to be applicable, and since the excitation was of a 0ND shape, only a mode with a 0ND character was thought to be able to be excited (for the inducer, a 0 ND was defined to be a shape where all 4 blades were in phase). However, when a forced response became necessary, it was discovered that blade 1ND and 2 ND modes could also be excited, and the conclusion reached was that the axial geometric nature of the blades actually caused "blade-like" behavior to dominate. The situation is somewhat clearer for impellers, where the shroud and the back-face are "disk-like" and the vanes are "blade-like."

**Fig. 3.23** Simple cylinder acousto-structural nodal diameter diagram

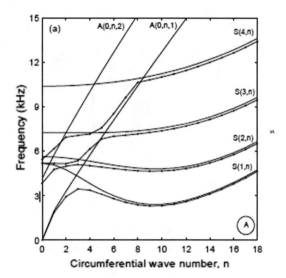

We have already mentioned modal analysis of impellers in Sect. 3.6.4.4 with regards to standing versus travelling waves. Another factor that must be included in impeller assessment is the excitability of modes as defined by the Tyler-Sofrin charts. The mechanism for the spatial-aliasing is not as clear for these structures as the pressure field from an adjacent flow-disturbance impinges not only on the discrete vanes but also on the continuous shroud. The continuity of this pressure-contact would seem to remove the aliasing effect. Nevertheless, engine programs do apply the charts to impellers in the same manner as other flow-path components, and there is some hardware evidence to support this approach [30, 31]. When generating these charts, the relative proximity of the impeller half-vanes to an excitation or responding piece of hardware need to be considered. A thorough research evaluation of the Tyler-Sofrin effect on impellers would be a valuable addition to the state-of-the-art.

### 3.6.9 Mitigating Resonance

It is frequently valuable to attempt to work to avoid a resonant situation as soon as it is identified via design techniques described below. As forced response analysis procedures become more cost-effective, this step will likely become more side-stepped, but for fundamental modes and for modes for which the forced response analysis does not come out well, the mitigation will still be necessary. The possible techniques include the following:

1. Redesign structure to move natural frequency away from excitation frequency.
2. Redesign flow-component configuration or engine speed to move excitation away from structural natural frequency;

**Fig. 3.24** LPSP turbine stator and 1st and 2nd stage turbine blades

3. Measure actual natural frequencies of hardware to reduce uncertainty.
4. Implement a significant damping or response-reduction mechanism.
5. Perform forced response to ensure acceptable ultimate and fatigue capability.

An excellent example of technique 1 was performed for NASA's LPSP LOX turbopump stator, built between 2011 and 2014. The stator was between two turbine blade stages with the same number of blades (Fig. 3.24), and the resonating mode (as flagged by a margin from the excitation of less than 5%, a small margin requirement since the engine was entirely for technology demonstration purposes) is a fundamental mode, as shown in the Campbell Diagram in Fig. 3.25. This combination would likely result in a high modal force (i.e., the mode would be driven efficiently by the excitation) so it was critical to move the stator out of resonance. A series of profile changes were attempted, but as the problematic mode moved out of resonance, a higher mode would move into range. Finally, a last-ditch effort threaded the needle between the excitations, and the design change was implemented into the final design (Fig. 3.26).

Technique 2 could be greatly effective, but engine system designers usually will not change flow parameters that control the engine speed, no matter how small; this lack of flexibility on their part may be due to a lack of understanding of structural dynamic issues. We will refer to technique 3 in the modal testing section, but the intent is to reduce the epistemic uncertainty and therefore reduce the margin requirement. Technique 4 will be discussed in Sect. 3.7, and technique 5 in Sect. 3.8.

## 3.7    Damping in Turbomachinery

As explained in Chap. 2, forced response depends on three main factors: (1) The excitation magnitude and shape; (2) the structural dynamic characteristics; and (3) damping. Unfortunately, characterizing the amount of damping in an operating turbopump component

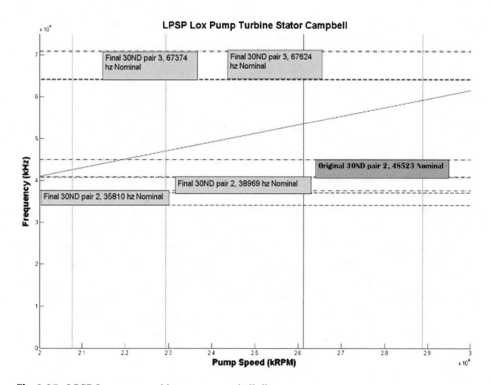

**Fig. 3.25** LPSP Lox pump turbine stator campbell diagram

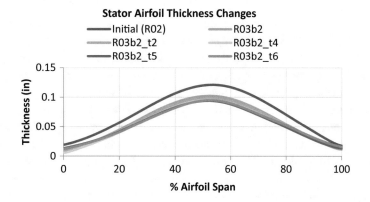

**Fig. 3.26** LPSP stator airfoil thickness optimization

still has significant uncertainties. Different sources of damping are material, mechanical, aero/hydrostatic, and aero/hydroelastic.

### 3.7.1  Damping in the Turbine Side

In turbines, the damping provided by the component moving through a quiescent gas is negligible (aerostatic damping). The damping provided by the movement through a flow (aeroelastic damping) is also small (~0.1–0.2% for jet engines [32]) and up to this time has been ignored in rocket engine turbine flow-path component analysis. However, with the advent of blisks, where the other sources of damping are also very small, the aeroelastic contribution may be important. There has, unfortunately, been only a small amount of research in this area specifically for rocket engine turbines (which yielded damping between 0.14% and 0.4%) [33], but the field has been extensively investigated for other turbine applications in the jet engine and power industries [34], so application to a rocket engine should not prove to be prohibitively difficult. In these other applications, the aeroelastic damping is a function of flow velocity and nodal diameter of the mode shape, and can actually become negative, or destabilizing, for certain combinations of these parameters, leading to the well-known flutter phenomena; this is unlikely for rocket turbine parameters, though.

Material damping is quite small, but not negligible. Measurements of metals have yielded damping values ranging between 0.05 and 0.15% for different modes [7] and even for Ceramic Matrix Composites, which were hoped to have higher material damping, of only 0.2% [5].

Although a well-designed turbine would never dwell on a primary mode of either a stator or blade, the required speed range of rocket engine turbomachinery will always take it through that mode transiently as the speed is ramped up. In addition, dwelling within the frequency margin of higher-order modes is frequently unavoidable. Therefore, a much higher amount of damping than supplied by the material itself has been required even if a forced response analysis is not performed. For the SSME, the initial requirement was 2%. For most turbine bladed-disk systems, this has been supplied by mechanical damping, which comes about both from the inherent design including some form of sliding or friction at the blade-disk interface, called the "fir tree," which provide at least 1% damping and by external, separate "dampers" inserted in some way into the structure. External dampers usually are inserted just below turbine blade platforms (called "under-platform dampers"), although there has been work attaching dampers to shrouds as well. The damping provided is a function of the normal load applied by the damper during operation, so the critical parameters are damper weight, rotational speed, and coefficient of friction. The shrouds themselves have also been designed to rub against each other, as in the S-3D turbines used in rocket engine for the 1960's era Saturn I launch vehicles (Fig. 3.27 [35]).

One of the problems with the friction technique is that the phenomena is completely non-linear because the amount of damping is proportional not just to the velocity, which is the case for a viscous damper, but it is also proportional to the magnitude of the normal force at the friction interface and the displacement. A predictive analytical capability therefore requires very advanced, state-of-the-art, and somewhat unvalidated techniques to predict the amount of damping provided, or an iterative, dedicated test program, which is very expensive.

The dedicated test program is generally performed using a high-speed spin-facility, sometimes called a "whirligig," which is necessary since the damper's function can only truly be measured while rotating in a configuration very close to actual operation. The bladed-disk is attached to a mechanical drive, and loading is applied using pressurized air jets, oil jets, magnets, or even acoustic speakers. The response of the blades is measured during the test using laser tip-timing or strain gages. As might be expected, the results vary over a non-trivial range, and determining the appropriate value from the data is problematic, i.e., merely using the average value may not capture the conservatism necessary for the engine program requirements. Unlike jet engine or other turbines, measuring the response during actual operation isn't possible for rocket engine turbomachines currently

**Fig. 3.27**  S-3D turbine shroud dampers

due to the extremely high temperatures and pressures and limited access for optical measurement devices, so determining a margin from actual test isn't possible. A proposal
that uses the combined effect of the damping statistics along with other non-deterministic
parameters in an overall non-deterministic analysis framework to back out the necessary
statistical measure for damping has been made by the author [36].

The response of the static hardware in turbines is frequently ignored as the mean load
is much less than rotating hardware so the alternating stress capability is much higher
(refer to Sect. 3.1). However, fatigue failures have occurred and if pieces are liberated
downstream, the consequences can be just as catastrophic as blade failures. The implementation of dampers into the static hardware, therefore, is rare, with response mitigation
relying entirely on frictional interfaces built into the design.

### 3.7.2  Damping in the Pump Side

The prediction and measurement of damping on pump-side components has not only the
same difficulties as on the turbine side but a number of additional ones as well. The operation of the components in a cryogenic liquid means that neither hydro-elastic damping,
which is a function of flow speed, or hydrostatic damping, which is the "viscous" damping
resulting from vibrating while submerged in a quiescent flow, are negligible. Additionally,
the presence of cavitation from inducers complicates both of the above damping phenomena, and tight tip clearances also affect damping. There has also been significantly less
testing and analysis documented in the literature for liquid pump components as the industrial application of turbines (or compressors in gas) are significantly larger than pumps.
The MSFC inducer project results have yielded preliminary estimates of the means for
the LH2 low pressure pump which are approximately 1% for the hydrostatic damping and
approximately 2% for the hydro-elastic damping, but the statistical spread is significant,
so the values used for a conservative analysis would be much less.

### 3.8  Forced Response Analysis

An accurate forced response analysis is intended to calculate the actual response of the
component in operation, and therefore would be the best solution to the question of sufficient structural dynamic capability. The state-of-the-art, though, is still not quite at the
level either in timeliness or accuracy to be used for every situation. This is particularly the
case for pump-side components, where the CFD for liquid propellants not only is not fully
developed but also because the fluid–structure interaction is significant, as discussed previously. For the turbine side though, enough development has taken place in recent years
in both speed of computation and in understanding various complicating effects that most
companies can undertake this analysis when needed. However, as complications in the

flow-path geometry increase, such as asymmetric inlet conditions or multiple stages, the analysis also become increasingly difficult. The GUIde Consortium in Turbomachinery Aeromechanics [37], an international group of multi-national corporations and NASA, sponsors extensive research specifically in this discipline to further the state-of-the-art in this complex field. This section will briefly review the required steps in a forced response analysis and mention the complicating effects; it is recommended that the references be studied to fully implement these.

## 3.8.1 Forced Response Analysis Procedure (Frequency Domain)

Forced response analysis for turbomachinery components generally is performed in the frequency domain since the loading is readily transformed to its spectral components. Although the technique is applied "under the hood" of commercial finite element codes, a thorough understanding of basic concepts as presented in Chap. 2 is critical for understanding the results.

To make forced response analysis of turbomachinery components tractable, modal superposition is generally applied. Component Mode Synthesis, which dominates launch vehicle structural dynamics, is generally not applied in turbomachinery. The reasons for this are that a turbopump component is built by a single vendor, unlike the many who would build a vehicle, and because cyclic symmetry helps with the model reduction needs. CMS has been applied in the Component Mode Mistuning method mentioned in Sect. 3.6.6, as a bladed-disk is suited for application of the method.

Determining the response of a single turbine bladed-disk to the excitation of a single inlet vane is the most straightforward type of response. The CFD is not at all trivial, with many considerations such as required length of the simulation and the required angle of spatial meshing. The CFD results can consist of either temporal histories of the forcing on each surface element, or frequency descriptions. If the results are temporal (transient), then they are generally run through a temporal Fourier decomposition to determine complex spectral components, as discussed in Sect. 3.4. In addition, as stated in Sect. 3.6.4, the analyst must choose how many harmonics of the primary excitation to keep, with 3 being the traditional rule-of-thumb but up to 7 necessary for sharp inlets like nozzles. The spectral component of interest is chosen based on the Campbell or Interference Diagram and is then spatially mapped onto the structural finite element mesh, which in general will be less refined than the CFD mesh. One complication that may be necessary to address is that many finite element codes cannot apply pressures directly onto a single, outer face of a solid element, in which case a non-structural surface shell element must be created for the pressure to be applied to. In addition, at present a forcing function can't be applied onto a structure within a liquid modelled using the modal-acoustic techniques described earlier, so additional code must be added to accomplish this loading.

The CFD pressures are dependent, of course, on the rotational speed that is used in the analysis. Frequently the structural dynamics analysis must be performed at a different speed, so the pressures will be incorrect. A rule-of-thumb to correct for this is that the magnitudes of the dynamic pressures can be scaled by the square of the speed. Although this is an approximation, it can serve adequately for the initial analysis, and another CFD at the new speed can be run if more accuracy is required.

An important parameter is how much of the turbine, both circumferentially and axially, has to be modelled. If the unsteady loads don't vary significantly from one blade to another for a given stage, then the circumferential angle subtended by the CFD can be less than 360°, and the response of a single blade in the row can be assumed to be the same for all the blades (ignoring mistuning). Since the number of excitation mechanisms is generally not equal to the number of responding entities (e.g., the number of stators isn't equal to the number of blades), the angle modelled has to be enough to capture this ratio reasonably well. For instance, if there are 30 stators and 43 blades, a section of 3 stators and 4 blades is adequate. This is called the "reference blade" approach. If the unsteady loads do change, generally due to multi-stage or asymmetric inlet effects, then modeling of the entire annulus will required.

If the turbopump is asymmetric, which is generally the case, then the CFD itself will also be. The extent of this can be determined using a two-dimensional temporal and spatial Fourier decomposition to determine the circumferential wave spectral composition, as shown in Fig. 3.28, from a study of asymmetric effects [20]. In this plot the typical frequency components are shown on the abscissa, and the number of circumferential waves are shown on the ordinate. For simple flow paths the circumferential wave numbers will be concentrated at the number of flow distortions and at that frequency, as discussed in Sect. 3.4, but as the geometry becomes more complicated, there will be additional content at other wave numbers, and at both the original frequency and other frequencies.

In any case, the first phase of an analysis will generally be a "simple" reference blade approach using a single temporal Fourier component. Even if the excitation frequency is not precisely at the natural frequency, general practice is to assume that it is, and to apply this complex Fourier component over a frequency range on either side of the resonance that either reflects uncertainty bounds (for both excitation and natural frequency) or the speed range. Special care must be made to specify a denser number of analysis frequency points near the natural frequency to ensure that the peak response is calculated.

Once the analysis is complete, post-processing must be performed to obtain the quantity of interest. For displacement, the process is straightforward; the frequency response results will be available for every node in either magnitude/phase or real/imaginary form for each coordinate direction, and we are generally interested in total magnitude, which can be obtained by root-sum-squaring the magnitudes for each direction. If we want dynamic stress, it is not generally known which location will be the peak stress, so a complete stress field (generally Von-Mises (Equivalent)) of the entire structure at the natural frequency should be obtained first. It is important to also obtain a frequency response

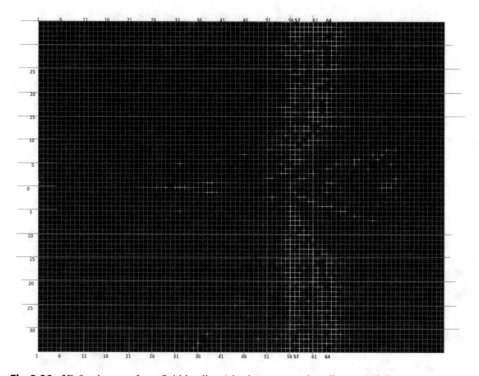

**Fig. 3.28** 2D fourier transform fluid loading (abscissa-temporal; ordinate-spatial)

plot to make sure the frequency resolution is enough for the peak value to be obtained as well as to gain the critical understanding of the response over the frequency range. However, finite element codes may not have the capability of calculating this value directly for Von-Mises stress, but only for normal and shear stresses at nodes, as shown in Fig. 3.29.

### 3.8.2 Effect of Mistuning Amplification Factor

For bladed-disks, one of the most important complicating effects is the amplification of the forced response due to mistuning, as discussed in Sect. 3.6.7 and shown in Fig. 3.17. Mistuning "localizes" the response energy towards one or two "rogue blades" and "amplifies" the response for almost all modes, and in the worst case up to (and maybe beyond) the Whitehead limit [38] of

$$MAF_{\max} = \frac{1 + \sqrt{N}}{2}$$

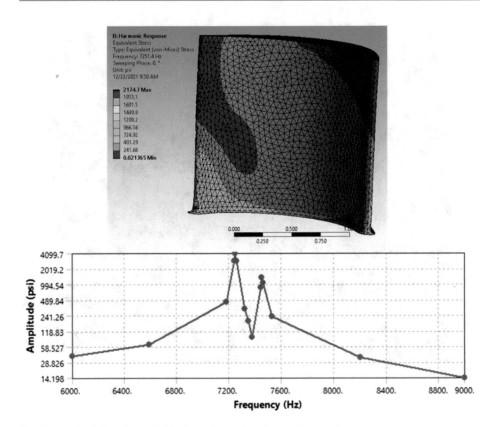

**Fig. 3.29** Von-Mises Stress field plot and associated normal stress frequency response

where $N$ is the number of blades, so this value can exceed 300%. Experimental evidence of *MAF* values this high are hard to find due the highly non-deterministic nature of mistuning. It is critical, however, that either a non-deterministic estimate of the maximum value of *MAF* be obtained during the design phase, where the actual mistuning is unknown, or that a deterministic value of *MAF* be obtained when mistuning parameters of specific bladed-disks are known as a result of modal testing being integrated into the production plan.

### 3.8.3 Non-adjacent Stages in Turbines, 360° Loading Variation, Transient Versus Frequency

Recent research has also shown that non-adjacent stages (i.e., different bladed-disk stages) can have a substantial impact on the loading on a particular turbine stage. It appears that the circumferential wave field has a number of components tied to the difference between non-adjacent counts. This is not intuitive, as there isn't relative rotation between the stages, but numerous studies and testing have shown the phenomena. Effects from both upstream and downstream stages are possible, although the upstream ones are larger [39].

Since cyclic-symmetric models are used for the structure, this variation in loading from blade-to-blade presents a significant problem. A difficult transform procedure that applies this 360° loading onto different sectors, which are then tracked, is necessary. MSFC has developed such a procedure, but it has not been published and still requires validation.

An implication of the above finding is that if a number of other wave numbers or frequency sidebands not at the primary frequency exist, this can seriously affect the accuracy of the frequency response methodology generally used. To examine this accuracy, a transient analysis must be performed to serve as the baseline, which is extremely time and memory intensive. Brown and Schmauch, and Gagne in a follow-on study, showed that frequency sidebands in the 0.25 N range (obtained if 4 revolutions of CFD are simulated) can have a very large impact on the response for certain spatial loadings, and that these are difficult to replicate in the frequency domain [20, 40].

### 3.8.4 Rules of Thumb for Amplitude of Dynamic Pressure

As reliable and timely CFD for the generation of unsteady pressure is still relatively new, rules of thumb for the magnitude of the dynamic forcing function have been developed for both the turbine side and the pump side. For the turbine side, this rule is that the unsteady pressure is approximately 5% of the mean pressure (one has to be careful not to refer to it as "dynamic" pressure vs "static" pressure, as there is flow in both cases). At this time, CFD techniques for gas turbines is refined and widely available enough that this value is only used as a conservative value for HCF capability calculations when the resonant frequency margin is adequate.

Pump-side unsteady CFD is still rare at this time, however. The rule of thumb, which has been verified experimentally repeatedly, states that the "dynamic pressure stress is 30% of the mean pressure stress" [41]. It is important to realize that the testing was performed by strain-gauging non-resonant inducer blades in a water flow facility, where the measured quasi-static stress should have the same ratio to the loading whether it is unsteady or steady. However, this "inverse force determination" methodology is extremely tricky because of resonant frequencies in the bandwidth of interest (see Chap. 4), and also cannot give a full-field description of the pressure field, so the modal force, which determines how effectively a mode is excited (see Chap. 2) can't be determined. For these reasons, the 30% rule is generally only used for non-resonant situations.

### 3.8.5    Effect of Dither (Unsteadiness of Operating Speed)

The rotational speed of a rocket engine turbomachine is not a designed characteristic of the engine, but rather a resultant of other input parameters, such as desired engine thrust, inlet pressure and temperature, and required outlet pressure and temperature. During the design process, the expected speed is uncertain due to the impossibility of precisely modeling the process causing the turbine rotation (epistemic uncertainty). But even after the machine has been built and tested, these parameters will have some inherent, aleatory variation, which will affect the resulting turbopump speed as well. This introduces the obvious question of how this variation will affect the assumption of resonant response in a frequency response analysis. The resonant assumption is clearly conservative, but if the dynamic stress from the resonant analysis violates a criterion, knowledge of the degree of conservatism is critical for acceptance of the design.

MSFC engineering investigated this topic extensively for the J-2X engine program in 2011 [42]. First, testing from both the Space Shuttle Main Engine and J2-X power-pack were examined, and as expected, a nontrivial amount of dither was found, with the speed following roughly Gaussian distributions with coefficients of variation at about 1% (Fig. 3.30). A methodology was then created to generate a SDOF system at the same natural frequency of a representative stator and apply a fully defined transient forcing function onto that stator using a Matlab$^\copyright$ ordinary differential equation solver to generate the "true" response. The damage was then calculated using the continuous version of Miner's Rule discussed in Sect. 3.3 and compared with the damage from a pure resonant excitation. The resulting "Dither Life Ratio", which is the factor of greater expected life for the true system compared to the resonant system, is a function of assumed damping and the statistical properties of the dither (Fig. 3.31), with the specific case being examined yielding a factor of 2.13, which is quite a large increase in expected life.

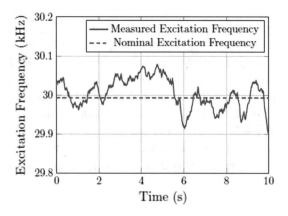

**Fig. 3.30**   10s history of J2-X powerpack excitation frequency

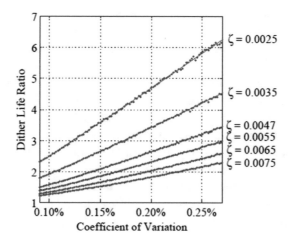

**Fig. 3.31**  Dither life ratio as function of CoV of speed and zeta

### 3.8.6    Cavitation and Other Non-integer Engine Order Loading

There are two notable forcing functions in turbomachinery that are not tied to integer multiples of the engine order. The first of these is cavitation from the pump inducer. As previously mentioned, the inducer ingests low-pressure propellant from the fuel tanks via a feedline duct, and mechanically adds pressure high enough for other pump mechanical elements (such as an impeller) to efficiently raise the pressures high enough for combustion. The limiting factor in how fast the inducer can spin to achieve this is the Cavitation phenomena, in which the vapor pressure of the fluid is not high enough to maintain contact with the inducer blades and bubbles form and collapse onto the blade. This is a very detailed subject, with extensive research in the literature, where the main structural concern is the erosion damage that these collapsing bubbles impart on the blade, which can be quite significant.

The oscillatory nature of cavitation and its ability to excite a blade natural frequency cannot be ignored either, though. There are two different types of Cavitation, called "low order cavitation" and "high order cavitation (HOC)", which, unlike all the excitations we've discussed so far, is tied to a non-integer multiple of the engine order [43]. HOC itself is composed of a rotating component which has been found between 5.4 and 5.7EO, which most efficiently excites 1ND modes, and an axial surge component found between 6.4 and 6.7EO, which similarly excites 0ND modes. The magnitude of cavitation is fairly substantial and is easily detected in acceleration measurements on the outside of the turbopump housing during hot-fire testing. This flowfield has been found to be a contributor to a number of failures in the rocket industry, including a Japanese H-2 inducer blade failure [44]. As part of the extensive inducer investigation performed at NASA/MSFC previously mentioned, a state-of-the-art forced response analysis was performed, but the details have not yet been published and are beyond the scope of this text. The cavitation bubble collapse additionally creates an acoustic wave which propagates upstream and so can affect hardware in the inlet duct. Analysis of these components will be discussed in Chap. 6.

Another non-integer excitation in turbomachinery is the well-documented and widely-observed phenomena of vortex-shedding limit cycle response. This has been determined to cause cracking in the SSME Low Pressure Oxidizer Turbopump inlet nozzle vanes, resulting in a life-limiting criteria. The phenomena have also been commonly observed in temperature probes in the engine system hardware (to be discussed in Chap. 6) and caused a major failure of the main injectors (see Chap. 4). The shedding frequency is associated with the flow speed and geometry, so is tied to engine speed only to the degree that flow speed is tied to it. The excitation frequency should be calculated by CFD, but it can be estimated using

$$ f_{shedding} = St * \frac{U}{L_{eff}} $$

where $St$ is the Strouhal number, approximately 0.2 for Reynolds Numbers between 500 and 500,000, $U$ is the flow velocity, and $L_{eff}$ is the effective length, e.g., diameter for a cylinder. The Reynolds number equals $U*L_{eff}$ divided by the kinematic viscosity of the fluid, which for LH2 is 1.3e-5 Pa-sec and LOX is 10e-5 Pa-sec. This phenomenon can be problematic because if the shedding frequency is just somewhat close to the structural fundamental natural frequency, then the non-resonant oscillatory forced response of the structure, which in general is very similar to the primary mode shape, will influence both the magnitude and frequency of the vortices, which in turn will then cause a feedback resonant response. The magnitude of this transverse response of the probe, which is independent of the initial conditions, can be substantial and can only be determined exactly using a nonlinear "limit-cycle" analysis, but it can be up to 1.7 times the diameter of the cylinder, causing huge stresses at the cylinder base [45, 46].

## 3.9    Other Turbopump Flow-Path Components

The large structural housings of both the pump and turbine sides do not require the extensive "component" structural dynamic analysis that we have been discussing up to this point. They will need to be included in the global engine "loads" model though, which will be discussed in Chap. 5. On the other hand, there are a number of subcomponents within the housing that are in the flow-path, and therefore experience significant dynamic loading requiring HCF or ultimate strength evaluation, and that have proven problematic in the past Although these components are stationary, which reduces the mean stress significantly compared to rotating hardware due to the lack of centrifugal load, they still have to withstand tremendous dynamic forcing functions emanating up or downstream from the bladed-disks, and failure can cause catastrophic consequences.

Within the turbine itself lie the stationary guide vanes, frequently called stators or inlet guide vanes. As discussed previously, the dynamic loading is harmonic, resulting from potential or wake disturbances. One key difference from bladed-disks is that stators and vanes are frequently fabricated in packets of 4–6 vanes; cyclic symmetry therefore doesn't really exist, along with all the implications discussed previously. Basic Campbell diagrams can therefore be applied successfully, therefore, and if there is a crossing, forced response analysis can be performed. A failure can be catastrophic, as shown in the SSME Unit 6–4 test-stand failure in 1996. Extensive hardware evidence backed up by analysis showed that the destruction of the high pressure fuel pump was due to a structural dynamic failure of either an inlet vane, a blade-outer-gas seal, or a combination of both, causing the vanes to rub the disk and a cascade of other failures.

Other stationary components include the turbine inlet bellows, struts and liners, turbine exit guide vanes, blade-outer-gas-seals, and pump diffuser vanes. These components undergo harmonic excitation from acoustic waves within the flow, as was discussed in Sect. 3.6.8, and from upstream random pressure waves emanating from the fuel pre-burner, which is a mini-combustion chamber creating the hot gas for driving the turbine.

HCF Cracking of the fuel inlet liner occurred during a SSME hot-fire test, so a thorough structural dynamic analysis was performed, yielding nodal diameter modes coupled with axial wave modes (Figs. 3.32 and 3.33). As with turbine bladed-disks, characterization of these shapes is critical for comparison with the acoustic nodal diameter and axial wave forms. After comparison with the frequency of these sources, as well as the random excitation coming from the fuel pre-burner, it did not appear that high resonant vibration was likely (more details on random response analysis will be presented in Chap. 4). Although this case did not pinpoint dynamics as a problem, it does show the importance of evaluating these types of components. Similar problems have arisen for other shell-like stationary turbopump structures such as the knife-edge seal and the fuel turnaround duct, where it is critical to identify both the axial and circumferential wave numbers of the modes to know which acoustic mode could be a source of excitation. Flow-induced vibration, a self-sustaining phenomenon similar to vortex-shedding, has also been found to cause cracking in knife-edge seals (as well as bellows commonly found in ducts and the fuel turbine inlet); that subject is quite extensive and out of the scope of this text, but the book by Blevins is valuable [47], as well as a number of NASA technical memos. Of course, not all turbomachinery will have these identical structures, but any element in the flow path should undergo a structural dynamic evaluation like the harmonic analysis previously discussed.

The main stationary components of concern on the pump side are the diffuser vanes. These vanes on the J-2X fuel turbopump showed a crossing on the Campbell of several modes with the 24EO and 36EO caused by upstream impeller vanes, and the ensuring

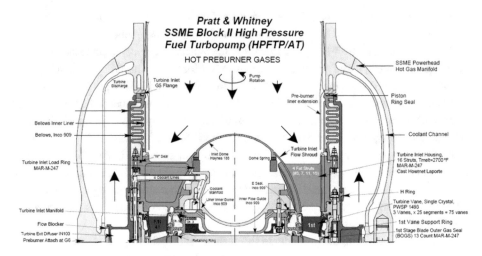

**Fig. 3.32** High pressure fuel turbopump turbine side

**Fig. 3.33** SSME turbine inlet liner 4th modal family

forced response analysis, which used a high fidelity CFD analysis to generate unsteady pressures, yielded dynamic stresses that exceeded HCF capability (Fig. 3.34). A life-limit was placed on the part, but the engine program was cancelled before that limit was reached during testing, so no further work was performed. Another diffuser issue was discovered on the SSME; during acceptance testing of an engine, a crack was found on the HPFTP pump 2–3 diffuser. It was believed that the global housing 2nd Nodal Diameter mode at 1192 Hz, which could be excited by 2EO (one of the always-present 1-6EO excitations), might cause excessive dynamic stress in the separate volute and diffuser part by enforcing high deflections at the interfacing boundary between the two parts (Fig. 3.35). When the cause of the excessive 2EO was eliminated, the cracks stopped as well.

It would be impossible to discuss the required dynamic analyses and problems discovered for various stationary hardware components during every engine program in this text. Nevertheless, hopefully enough material has been presented to enable the analyst to identify which turbopump structures require examination and how those analyses would be performed.

**Fig. 3.34**  J-2X FTP pump diffuser

**Fig. 3.35**  SSME HPFTP
pump housing and diffuser

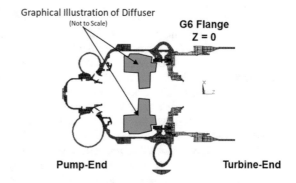

## References

1. Jakobsen J, Liquid Rocket Engine Turbopump Inducers', NASA SP 8052, 1971.
2. Zoladz TF, Lunde K, Mitchell W, SSME Investment in Turbomachinery inducer-Impeller Design Tools, Methodology and Test, 57th JANNAF Joint Propulsion Meeting, 2010.
3. Atkinson RJ, Winkworth WJ, Norris GM, Behaviour of Skin Fatigue Cracks at the Corners of Windows in a Comet I Fuselage, Ministry of Aviation via Her Majesty's Stationery Office, 1962.
4. Campbell W, The Protection of Steam Turbine Disk Wheels From Axial Vibration', ASME Spring Meeting, 1924.
5. Min JB, Harris DL, Advances in Ceramic Matrix Composite Blade Damping Characteristics for Aerospace Turbomachinery Applications', 52nd AIAA Structures, Structural Dynamics, and Materials Conference, 4–7 April 2011, Denver, Colorado.
6. Xu Li , Xueyi Li , Changyou Li , David He, Fatigue life prognostic for medium-carbon steel based S-N curve computation and deep autoencoder Vibroengineering PROCEDIA, Vol. 20, 2018, p. 64–69. https://doi.org/10.21595/vp.2018.20146
7. Brown, AM, DeHaye M, DeLessio S, Probabilistic Methods to Determine Resonance Risk and Appropriate Damping for Rocket Engine Turbine Blades', AIAA Journal of Propulsion and Power, 2013. https://doi.org/10.2514/1.B34834
8. Marcu B, Tran K, Dorney D, Schmauch P, Turbine Design and Analysis for the J-2X Engine Turbopumps', 44th AIAA Joint Propulsion Conference & Exhibit, July 2008.
9. Brown AM, Comprehensive Structural Dynamic Analysis of the SSME/AT Fuel Pump 1st Stage Turbine Blade, NASA TM-1998-208594.
10. Saad Y (2011) Numerical Methods for Large Eigenvalue Problems, ISBN 9781611970722.
11. Campbell, W., Protection of Steam Turbine Disk Wheels from Axial Vibration, Schenectady, N.Y., General Electric Co.
12. Coleman, HW, Steele WG (2009), Experimentation, Validation, and Uncertainty Analysis for Engineers, Third edition, John Wiley & Sons, Hoboken, NJ.
13. Castanier MP, Pierre C, Modeling and Analysis of Mistuned Bladed Disk Vibration: Status and Emerging Directions, Journal of Propulsion and Power, Vol. 22, No. 2, March-April 2006.
14. Singh MP, Vargo, JJ, Schiffer DM, Dello JD, Safe Diagram - A Design And Reliability Tool For Turbine Blading, 1988 Turbomachinery and Pump Symposium, Texas A&M University Turbomachinery Laboratories.
15. Tyler, JM, Sofrin, TG, Axial Flow Compressor Noise Studies," SAE Transactions, Vol. 70, pp. 309–332, 1962.
16. Figaschewsky F, Kühhorn A, Beirow B, Giersch T, Schrape S, Analysis of mistuned forced response in an axial high-pressure compressor rig with focus on Tyler–Sofrin modes, The Aeronautical Journal, 123(1261), 2019, 356–377. https://doi.org/10.1017/aer.2018.163
17. Kushner F, Rotating Component Modal Analysis and Resonance Avoidance Recommendations', Tutorial, Proceedings of the 33rd Turbomachinery Symposium, Turbomachinery Laboratory, Texas A&M University, College Station, TX, 2004
18. Jay RL, MacBain JC, Burns DW, Structural Response Due to Blade Vane Interaction, Journal of Engineering for Gas Turbine and Power, January 1984, Vol. 106, pp. 50–56.
19. Hegde S, Zori L, Camregher R, Kielb R, Separation of Wake and Potential Field Excitations in an Embedded Compressor Rotor: Impact of wave reflections and mistuning on Forced Response', AIAA Scitech Forum, 11–15 January 2021, AIAA 2021-0265
20. Brown AM, Schmauch P, Characterization of Deficiencies in the Frequency Domain Forced Response Analysis Technique for Turbine Bladed-Disks', 53rd AIAA Structures, Structural Dynamics and Materials Conference, 23–26 April 2012. https://doi.org/10.2514/6.2012-1397

21. Yang MT, Griffin JH, A Reduced Order Model of Mistuning Using a Subset of Nominal System Modes, Journal of Engineering for Gas Turbines and Power, Vol. 123, No. 4, 2001, pp. 893–900.
22. Kielb RE, Kaza K, Effects of Structural Coupling on Mistuned Cascade Flutter and Response', J. Engineering for Gas Turbines and Power, Vol. 106, pp. 17–24, January 1984
23. Beck JA, Brown JM, Cross CJ, Slater JC, Geometric Mistuning Reduced Order Models for Integrally Bladed Rotors with Mistuned Disk-Blade Boundaries', ASME Turbo Expo 2013, GT2013-94361
24. Lim SH, Bladh R, Castanier MP, Pierre C, Compact, Generalized Component Mode Mistuning Representation for Modeling Bladed Disk Vibration', AIAA Journal, Vol. 45, No. 9, September 2007
25. Feiner DM, Griffin JH, Mistuning Identification of Bladed Disks Using a Fundamental Mistuning Model—Part II: Application, Journal of Turbomachinery, 126(1), 2004, pp. 159–165.
26. Brown AM, DeLessio JL, Jacobs PW, Test-Analysis Modal Correlation of Rocket Engine Structures in Liquid Hydrogen—Phase II, IMAC- XXXVIII Conference and Exposition on Structural Dynamics, Houston, Texas, February 10–13, 2020, paper 8732.
27. Lindholm U, et al., Elastic Vibration Characteristics of Cantilever Plates in Water, Journal of Ship Research, June 1965, pp. 11–36
28. Xiu H, Davis RB, Romeo RC, Edge clearance effects on the added mass and damping of beams submerged in viscous fluids', Journal of Fluids and Structures, 2018. 10.1016.
29. Davis RB, Virgin LN, Brown AM, Cylindrical Shell Submerged in Bounded Acoustic Media: A Modal Approach', AIAA Journal, Vol. 46, No. 3, March 2008. https://doi.org/10.2514/1.31706
30. Internal letter, Failure of SSME Impeller, Gary Davis, Rocketdyne, 1986.
31. RS-25 High Pressure Fuel Turbopump First Stage Impeller Life Limit, RS25-5.6-032-0000-0001, Al High, Unpublished, March 2016.
32. Hall KC, Kielb R, Thomas JP, Unsteady Aerodynamics, Aeroacoustics, and Aeroelasticity of Turbomachines, Springer Press, 2006.
33. Besem FM, Kielb RE, Galpin P, Zori L, Key NL, Mistuned Forced Response Predictions of an Embedded Rotor in a Multistage Compressor, Journal of Turbomachinery, 138(6)/061003, 2016.
34. Smith TE, Aeroelastic Stability Analysis of a High-Energy Turbine Blade, 26th AIAA Joint Propulsion Conference, July 16–18, 1990, Orlando, FL, AIAA-90-2351.
35. http://heroicrelics.org/ussrc/engines-s-3d-turbine/dsc80316.jpg.html
36. Brown AM, DeHaye M, DeLessio S, Application of Probabilistic Methods to Assess Risk due to Resonance in the Design of J-2X Rocket Engine Turbine Blades, 52nd AIAA/ASME/ASCE/AHS/ASC Structures, Structural Dynamics and Materials Conference, Denver, Co., April 4–7, 2011, AIAA#944758.
37. The GUIde 7 Consortium Center for Aeroelasticity, https://aeromech.pratt.duke.edu/
38. Whitehead D, Effect of mistuning on the vibration of turbomachine blades induced by wakes', Journal of Mechanical Engineering Science, 8(1), 1966, pp. 15–21.
39. Hegde S, Multi—Row Aeromechanical and Aeroelastic Aspects f Embedded Gas Turbine Compressor Rotors, Ph.D Dissertation, Duke University, 2021.
40. Gagne, AA, Effects of Asymmetry and other Non-Standard Excitations on Structural Dynamic Forced Response Analysis of Turbomachinery Flow-Path Components', Master's Thesis, KTH University, Stockholm, Sweden, 2014.
41. Herda DA, Gross RS, High Pressure Oxidizer Turbopump (HPOTP) Inducer Dynamic Environment', NASA-TP-3589, November 1995.
42. Brown AM, Davis RB, DeHaye MK, Implementation of Speed Variation in the Structural Dynamic Assessment of Turbomachinery Flow Path Components', Journal of Engineering for Gas Turbines and Power, October 2013, Vol. 135/102503. https://doi.org/10.1115/1.4024960

43. Subbaraman M, Burton K, Cavitation-Induced Vibrations in Turbomachinery: Water Model Exploration," Fifth International Symposium on Cavitation (CAV2003), Osaka, Japan, November 1–4 2003.
44. Matsuyama K, Ito T, et al., H-IIA Rocket Engine Development, Mitsubishi Heavy Industries, Ltd. Technical Review Vol. 39, No. 2 (Jun. 2002).
45. Iwan WD, Blevins RD, A Model for Vortex Induced Oscillation of Structures, ASME Journal of Applied Mechanics, September 1974, pp. 581–586.
46. Khalak A, Williamson C, Motions, Forces and mode transitions in vortex-induced vibrations at low mass-damping. Journal of Fluids and Structures, Elsevier, 1999, 13 (7–8), pp. 813–851. https://doi.org/10.1006/jfls.1999.0236
47. Blevins, RD (1977), Flow-Induced Vibration, Van Nostrand, Reinhold, NY.

# Structural Dynamics of LRE Combustion Devices

**4**

## 4.1 Introduction

Components falling into the combustion devices category are the second major area of structural dynamic analysis for LRE's. There are only a few components in the category, consisting of the following:

1. Gas Generators/Pre-burners
2. Thrust Chamber Assembly (TCA), including injectors
3. Nozzle, including attached manifolds.

As these components are the energy-generating components of an LRE, though, failure of any of them is frequently catastrophic. Indeed, most of the early SSME development failures were of combustion devices, some of which were structural dynamic in nature, as mentioned in Chap. 1. In addition, the nozzle is either the structural backbone of the engine or such a dominant mass that engine system loads (Chap. 5) are always dependent on nozzle response.

## 4.2 Random Vibration Assessment of Gas Generator

The Gas Generator (GG) is small device that generally uses a solid propellant to generate hot gas to drive the turbopump turbines. This gas passes through and is exhausted from the

**Supplementary Information** The online version contains supplementary material available at https://doi.org/10.1007/978-3-031-18207-5_4.

**Fig. 4.1**  F-1 gas generator [1]

turbines into the ambient environment, thereby limiting the engine efficiency. This process identifies a major characteristic of these types of engines, so they're even called "gas generator cycle" engines. The device can take on many geometries but essentially consists of a container for the solid propellant and a small thrust chamber to converge and then diverge the combustion process (Fig. 4.1). A basic structural dynamic assessment of gas generators consists of a random vibration spectrum applied as a base acceleration excitation to the interface of the GG with the rest of the engine, which is assumed to be a fixed boundary condition. This process is described in Sect. 2.5.8.

As described in Sect. 2.6, PSD's of the alternating stresses are calculated and a 3-sigma value compared with the material capability. As the GG itself is the source of random vibration, and the base is not a fixed boundary condition but rather has the flexibility of the interface, this analysis is inherently flawed. Nevertheless, as will be further discussed in Chap. 5, quantification of the random forces coming from the GG are virtually impossible to obtain, so the only available option is to use acceleration environments that have been obtained by measurement of similar previous engines, or by measurement of the GG environment during post-design testing of the specific engine under consideration. Alternatively, the GG can be assessed similarly to many of the components mounted on the skin of a launch vehicle using the shock response spectra (SRS). This method is explained in Sect. 2.3.6, and it will yield the worst possible response for a given environment and assumed damping value. As the GG generally is built with a fairly substantial cross-section to contain the combustion process, it will generally pass the SRS analysis.

## 4.3 Structural Dynamic Assessment of Main Injectors

The main injectors are used to inject small amounts of hot oxidizer gas and fuel gas into the main combustion chamber, where they essentially self-ignite, although the process is assisted with an "augmented spark igniter," (Fig. 4.2). The large number of cylindrical injector elements (600 in the SSME) are of course subject to a high level of random vibration from the combustion process, but an additional excitation is caused by vortex shedding, as discussed in Sect. 3.8.6. This process destroyed a number of engines in the early 1980's (Fig. 4.3) until flow shields were implemented to reduce the flow velocity and detune the injectors from the excitation. As this solution was not optimal for performance, Rocketdyne and MSFC performed and documented a detailed test analysis program more clearly identifying the cause of the problem and preferable methods for detuning the injectors, such as implementing a stronger injector design [2]. It is clear, therefore, that the design of any rocket engine must assess this condition for safe operation.

## 4.4 Structural Dynamic Assessment of Nozzles

The nozzle of a rocket engine is surely its most recognizable feature. There are several different types of nozzle designs to satisfy the constraints of phase of launch (i.e., boost phase from sea level, or high-altitude for an upper stage), weight, reusability, and volume. All of the designs must somehow withstand the incredibly high temperatures reached within the exhaust bell (6000 °F for the SSME). The solutions to this problem seem to me to be "Rube-Goldberg" contraptions that somehow function and extract every bit of energy from the system. One method is to have either all or a section of the nozzle be

**Fig. 4.2** SSME main injector [3]

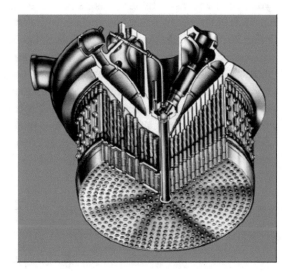

Test 750148 S/N Engine 0110

**Fig. 4.3**  Main injector failure SSME

"regeneratively cooled"; this amazing design uses hundreds of variable-diameter tubes brazed together through which the liquid hydrogen flows as the structure itself, thereby not only keeping the nozzle from melting but serving a critical function in heating up the hydrogen to a more useful temperature. Other techniques use either composite or metallic materials as an ablative inner layer, which will literally burn away in a controllable fashion during its single-flight profile. The altitude of operation is also critical in that the optimal nozzle area ratio (exit plane diameter/throat diameter) increases significantly with lower ambient pressures to maximize the thrust efficiency; however, launch vehicle upper stages may not have the volume to contain these huge nozzles. The RL-10, an almost 50-year-old engine whose variants are still being used, solves this problem with a movable nozzle extension that is stowed encircling the main engine section during first stage ascent, then when the first stage is separated it translates down using screw drives into operational position mating with the upper part of the nozzle—another incredible solution!

### 4.4.1  Modal Analysis of Nozzles

As with the other LRE components we've discussed, a modal analysis is a critical first step for nozzles and we need to consider the character of the excitation to determine which modes will be important. Almost all of the dynamic excitation on the nozzle is called "self-induced random", in that the excitation is generated with the thrust chamber assembly itself, specifically from the combustion chamber. This is not generally of sufficient amplitude at any one frequency to drive a resonant response. However, the nozzle is so large that it will drive the SD response of the entire engine, so a reasonably accurate model of the nozzle is critical.

Unfortunately, for the regeneratively cooled tube-walled nozzles described above, the finite element modeling itself is quite difficult. In the past, several approximating methods were used rather than explicitly modelling each of the hundreds of tubes that are brazed together, where the braze properties are widely variable. The task is made even more difficult as ducting to supply the hydrogen is generally mounted on the nozzle, so those ducts must be included in the model. More recent engine programs have accepted the difficulty and attempted to generate these enormous and tedious models. Even with this high-fidelity capability, model calibration with modal test is required (to be discussed in Chap. 7) because of the variable braze properties. As the fabrication of these tube-wall nozzles are incredibly expensive and time-consuming (each of the SSME nozzles cost $20 M and took 4 years to build), they are being slowly replaced with a "channel-wall" concept, in which channels for the liquid hydrogen are cut into a continuous nozzle-shaped plate which is then encased in a jacket, enabling the flowing LH2 to be routed throughout the nozzle and back to the engine circuit.

Once an adequate model is created, the modal analysis will yield modes that fall roughly into two families, bending and nodal diameter. The bending modes bend about the throat of the nozzle, the most flexible location, and the nodal diameter modes circumscribe waves in the circumferential direction. These are shown in a model of the entire Fastrac engine, developed by MSFC circa 2000. The nozzle for this engine is fabricated entirely with composites, a graphite epoxy base and a silica-phenolic overwrap. Although more tractable than tube-wall nozzles, this design had its own modeling challenges representing the composite layers [4], especially since there was a non-trivial variation in thickness. In this model, all the engine components except for the nozzle are represented by beam finite elements, and the nozzle is created with plate elements calibrated to the composite structure, so the nozzle displacement can be seen clearly. Figure 4.4 shows the first bending mode at 123.2 Hz, and Fig. 4.5 shows the first nodal diameter model at 147.4 Hz, which is a 3ND, not a one or two ND as would be first guessed. For this case and many of other nozzle configurations, this mode can be shown to have the lowest ratio of modal potential energy to modal kinetic energy for the nodal diameter modes, which is the basis of an alternate method of fundamental eigenvalue extraction, the Rayleigh–Ritz method.

Another complicating factor for this nozzle was that the composite nozzle anisotropic stiffness properties are highly temperature dependent. Since these temperatures change significantly during operation, this means that the natural frequencies change during operation as well. Since these frequencies are critical for loads calculation (to be discussed

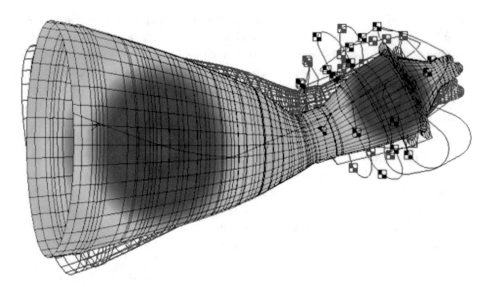

**Fig. 4.4** Fastrac first bending mode (undeformed shape superimposed)

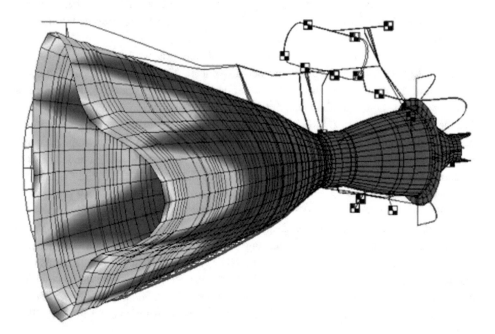

**Fig. 4.5** Fastrac 3ND mode

in Chap. 5), accurate identification of these frequencies is required. A complex hot-fire test/analysis correlation program was undertaken to accomplish this goal [5].

## 4.4.2   Dynamic Assessment of Nozzles

Depending on the particular engine design, evaluation of the SD capability of the nozzle itself, as well as the myriad ducting and manifolds attached, can be accomplished in a similar manner to the GG; the random environment is applied at an assumed fixed base (the gimbal), a base excitation enforced acceleration analysis performed, and a PSD obtained and margins calculated. If the turbopumps are attached to the nozzle via a stiff duct or a rigid powerhead, then a substantial harmonic forcing function at the rotational speed (and some multiples of that) will also propagate into the nozzle and need to be applied at the interface location in the same manner. Acoustic excitation rebounding from the launch pad and internal pressure oscillations during "main-stage operation" (i.e., after the transient start-up) can also be large excitation mechanisms and will be discussed in this section.

For a base drive analysis, the modal participation factors and effective mass for each mode become valuable (see Sect. 2.5.8.1). These calculations are usually performed by default by commercial finite element codes when a base drive is performed. For example, Fig. 4.6 shows typical nozzle first bending and second bending modes for a base excitation in the Y direction. As required, all DOF's at the base (the face on the left side of the picture at the combustion chamber/engine interface) are fixed.

The participation factor and effective mass table is retrieved from the output and is shown in Table 4.1. The only modes with appreciable participation factor magnitudes are the first and second bending in the X and Y directions, modes 1, 2 10, and 11. The effective masses reflect this distribution as well and show that 88.5% of the mass of the structure is accounted for by these four modes. The fact that the nodal diameter modes, such as mode 3 (Fig. 4.7), have negligible magnitude is critical, indicating that base excitation generated by mechanical vibrations transmitted into the nozzle from the combustion process or from elsewhere in the vehicle cannot excite these modes; therefore, the locations with high modal stress are not in the large bell portion but instead in the throat section that is highly stressed in the bending modes. This phenomena is discussed in detail in a highly-recommended 2014 paper on dynamic loads in nozzles by Pray and Baker [6]. The stresses from the bending modes, which are primarily in the throat region, are generally calculated as part of the overall engine system model, to be discussed in Chap. 5.

There are several excitation mechanisms that do excite the nodal diameter modes, though. Two of these are acoustic loads rebounding off the launch pad and internal pressure oscillations during main stage nozzle operation (i.e., after the start-up transients), also examined in detail by Pray and Baker. The main variables studied in the paper are

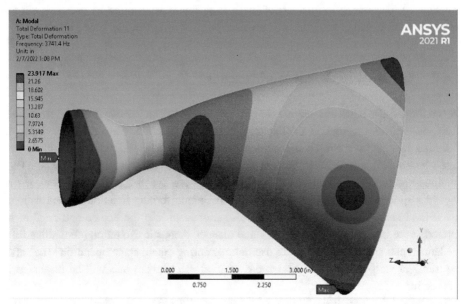

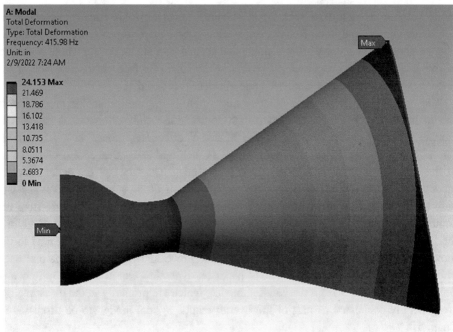

**Fig. 4.6** First and second nozzle bending modes

**Table 4.1**  Participation factor partial output from ANSYS

| Mode | Frequency | Participation factor | Ratio PF to total | Effective mass | Ratio effective mass to total mass |
|------|-----------|---------------------|-------------------|----------------|-----------------------------------|
| 1 | 415.983 | −1.0946E−02 | 0.200992 | 1.1982E−04 | 2.5378E−02 |
| 2 | 415.997 | 5.4461E−02 | 1 | 2.9660E−03 | 6.2819E−01 |
| 3 | 684.072 | 1.0082E−06 | 0.000019 | 1.0165E−12 | 2.1530E−10 |
| 4 | 684.086 | 2.3579E−06 | 0.000043 | 5.5596E−12 | 1.1775E−09 |
| 5 | 872.226 | 3.0354E−08 | 0.000001 | 9.2137E−16 | 1.9514E−13 |
| 6 | 1516.42 | −7.1670E−09 | 0 | 5.1367E−17 | 1.0879E−14 |
| 7 | 1516.44 | 1.2797E−08 | 0 | 1.6377E−16 | 3.4687E−14 |
| 8 | 2593.92 | −1.8146E−08 | 0 | 3.2926E−16 | 6.9737E−14 |
| 9 | 2594.07 | −4.5944E−09 | 0 | 2.1109E−17 | 4.4707E−15 |
| 10 | 3741.28 | −3.1917E−02 | 0.586053 | 1.0187E−03 | 2.1576E−01 |
| 11 | 3741.38 | 8.7888E−03 | 0.161378 | 7.7244E−05 | 1.6360E−02 |
| 12 | 3754.57 | −1.2982E−05 | 0.000238 | 1.6854E−10 | 3.5697E−08 |
| 13 | 3754.62 | 1.0414E−04 | 0.001912 | 1.0845E−08 | 2.2968E−06 |
| 14 | 3849.54 | −1.9992E−05 | 0.000367 | 3.9967E−10 | 8.4649E−08 |
| 15 | 3955.74 | 4.6778E−08 | 0.000001 | 2.1882E−15 | 4.6345E−13 |
| 16 | 3955.94 | 1.2690E−07 | 0.000002 | 1.6104E−14 | 3.4107E−12 |
| 17 | 4511.75 | 1.5315E−08 | 0 | 2.3454E−16 | 4.9675E−14 |
| 18 | 4511.95 | 4.1992E−07 | 0.000008 | 1.7633E−13 | 3.7347E−11 |
| 19 | 5605.46 | 4.4224E−09 | 0 | 1.9557E−17 | 4.1422E−15 |
| 20 | 5605.97 | −1.9531E−08 | 0 | 3.8145E−16 | 8.0791E−14 |
| Sum | | | | 4.1818E−03 | 0.885691 |

the magnitudes, temporal frequency spectra, and spatial distribution. The magnitudes can really only be obtained from measurements and scaling from similar engines. The spatial distributions are generated by examining statistical spatial correlation of the pressure field, where locations that are highly correlated are considered to be in-phase and so define a wave number. This wave number must be close to the nodal-diameter wave number, as discussed in the last chapter, for efficient excitation of that mode. If there is a match, the software code VA-One efficiently applies the statistically defined pressure field onto the structural dynamic model generated by NASTRAN. A sample case using the SSME is presented and results in non-trivial stresses that could be used to optimize the nozzle design. Similarly, upper stage engine modes, both bending and nodal diameter, can be excited by acoustics of the interstage during boost phase, while the engine itself is not operating; the RL10 engine used in NASA's Space Launch System had to pass a separate acoustic qualification test for this reason.

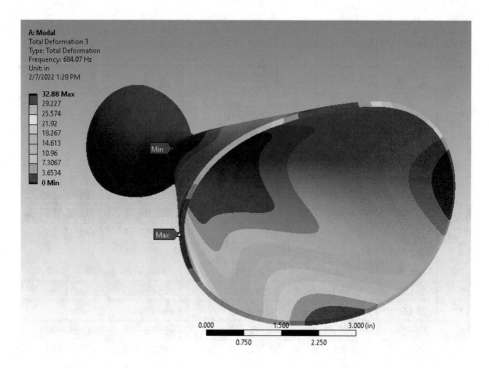

**Fig. 4.7** Two-nodal-diameter mode of typical nozzle

In addition, these pressure loads can combine with the base drive loads to cause potential buckling of the nozzle [7]. Nozzle buckling is characterized by massive (and destructive) deformation into an oval shape; this shape is a stable configuration of a conically shaped structure, just like a buckled beam curved shape is a stable configuration of an axially loaded beam. Several engine programs have purposely initiated nozzle buckling during hot-fire to assess the result, and the magnitude of the deformation seen in videos from these tests is astounding (unfortunately, the videos are not available for public release).

The response to these forcing functions is made more problematic by the significant lowering of Young's Modulus and loss of capability, both high cycle fatigue and ultimate strength, at high temperatures. These effects are shown in Fig. 4.8a and b respectively, for Haynes 230 [8], a typical high temperature material that was to be used for the J2-X nozzle extension which would reach a peak value of 2000 °F near the aft end. This is especially an issue for the composites and special metals used for uncooled nozzle extensions, where the highest temperatures are reached. As peak dynamic stresses for nodal diameter modes occur in the large nozzle extension, addressing these stresses with a thicker design will incur a non-trivial weight penalty, so accurate analytical predictions are critical for overall mission viability as well as safety.

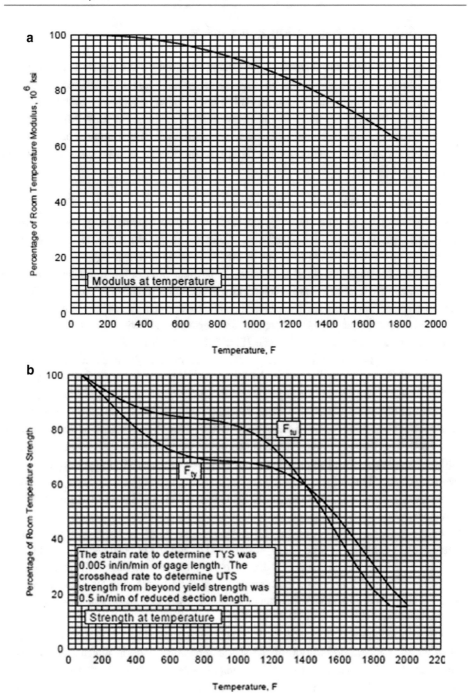

**Fig. 4.8   a** Haynes 230 E versus Temperature; **b** Ultimate Tensile Strength versus Temperature

### 4.4.3  Side Loads Separation Phenomena

The most significant nodal-diameter type forcing function on the nozzle is a result of a complex fluid–structure interaction phenomena generically called "Side Loads". As with mistuning in turbopumps, this phenomenon has been studied extensively, both by industry and academia, but a vast majority of the studies have focused on the fluid dynamics rather than the structure. Side loads have caused catastrophic total engine failures during testing of the J2-S second stage engine for the Saturn V, the SSME, the European Vulcain, and the Japanese LE-7A [9, 10].

#### 4.4.3.1  Start-Up/Shut-Down Side Loads

The mechanism for engine start-up side loads is shown schematically in Fig. 4.9. In Fig. 4.9(a) a graph of the internal pressure versus the longitudinal nozzle location as the nozzle fills up shows that the pressure in the combustion chamber $p_{wall}$ is much higher than ambient pressure outside of the bell $p_{amb}$, but that $p_{wall}$ drops significantly through the throat and then as it expands supersonically inside the bell. At some point, the pressure reaches a critical ratio to $p_{amb}$ for which the shock wave can no longer maintain contact with the inner wall due to this differential and separates, allowing the ambient pressure to flow in from outside and fill the gap (Fig. 4.9b). As with every real system, though, this separation is not axisymmetric due to flow and structural circumferential distortions, so there will be separation at one angular location while there is no separation at another. A little further downstream, the separation will finally occur at all angular locations and the pressure differential imparts a huge transverse transient load (called the "side load") equal to the integrated pressure differential over the sliver-shaped region of the differential (Fig. 4.9d). This short-duration load is called quasi-static to delineate it from steady-state loading described below. Initial estimates of this load in the J2-X engine during were on the order of a million lb!

This value is far too conservative, though, because it was based on an approximate technique called the "skew-plane method," developed by NASA during the shuttle program, in which the initial location of separation is assumed to occur at the maximum pressure ratio thought possible, $p_{wall}/p_{ambient} = 0.5$, and the opposite separation location is assumed to be at $p_{wall}/p_{ambient} = 0.25$, the minimum pressure ratio for separation. This method is still used to generate a conservative estimate of the side load for quasi-static buckling analysis during engine start-up or shut-down, when this load can cause the nozzle to massively (and destructively) deform into an oval shape.

It is well recognized, though, that the skew-plane approximation does not really correlate with a realistic description of the shock separation pattern, which is much more complicated and broken up into smaller, conical patterns called "Free" and "Restricted Shock Separation" [11]. A number of more refined techniques have been applied to generate this value, including a Monte Carlo technique used to assess the SSSME after the major failure in 1979. For that analysis, the random characteristic of the forcing function

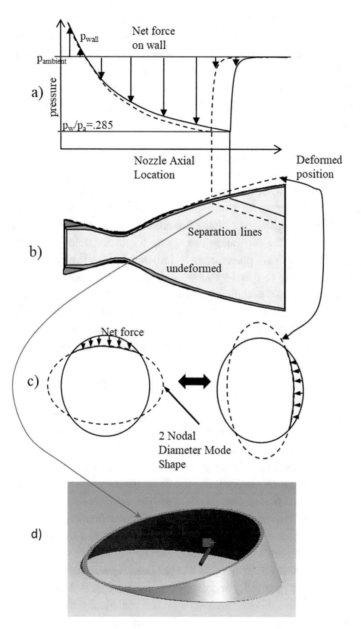

**Fig. 4.9  a**) Internal pressure versus axial location; **b**) schematic of differential pressure field; **c**) initiation of limit cycle oscillation; **d**) pressure differential region

seen in J2-S engine data is used to create four random parameters: the amplitude of the side load, the direction, a scaling factor from the J-2S to the SSME, and temporal variation. Even though these effects are transient during engine start-up and shut down, when the effect is just reversed in time, the critical cases of nozzle buckling, and the system effects of the huge transient transverse load being absorbed by the thrust vector control actuators and other engine system hardware require that the structural dynamics of the nozzle be accurately modelled.

### 4.4.3.2  Steady State Fluid Structure Interaction Side Loads

In addition, for "overexpanded nozzles", which is when the ambient pressure is greater than the design exit pressure, the engine can operate in a steady-state separated condition. In this scenario, which exists for many sea-level operating engines as well as high-altitude engines for which the operating altitude conditions cannot be replicated (a difficult and expensive proposition), the separation can set up a self-sustaining limit cycle oscillation as shown in Fig. 4.8c. This phenomena is initiated by the first asymmetric side load, which causes the nozzle to quasi-statically deform into an oval almost exactly (but not quite) the same as the two nodal-diameter mode shape. The magnitude of the structural deformation is so large that it actually causes the wall pressure at the outward-bowing location 90° away to drop, which then preferentially causes the flow to separate at that location, causing a side load there. This cycle will continue, only limited by the structural stiffness and damping.

This feedback has been observed in both in a sub-scale nozzle experiment and the full scale hot-firing of the Fastrac engine at MSFC [9], and for the SSME. A video of SSME startup (Online Resource 10) shows the restricted shock separations, indicated by the "tee-pee" shapes on the inside of the nozzle, and the responding two nodal diameter mode. Although the startup transient mechanism does not set up the limit cycle, the magnitude of the deformation is large enough to be easily seen with the camera, and clearly could be a structural concern. For the Fastrac hot-fire (Fig. 4.10), photos show the flow separation. Special heat-protected strain gauges were also mounted and the resulting temporal and spatial response also show the two nodal diameter response (Fig. 4.11), which as mentioned previously would not normally be excited by the transverse random based excitation due to the low participation factor.

Although an observed FSI limit cycle has not been reported in the literature elsewhere, it is certainly a distinct possibility and requires assessment. A technique incorporating this fluid/structure dynamic limit cycle along with the statistically characterized quasi-static load using the SSME method was used for the design of the J2-X engine that was to be used for the cancelled NASA Constellation program of the mid-2000's [12]. The technique was not really validated, though, but a more recent development of a fluid/structure capability between a high fidelity CFD code, LOCI, and the structural dynamics code NASTRAN was applied to this problem and was also able to show the inducing of the

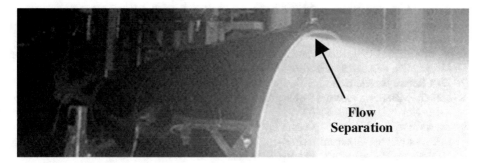

**Fig. 4.10**  Flow separation during hot-fire test of FASTRAC nozzle

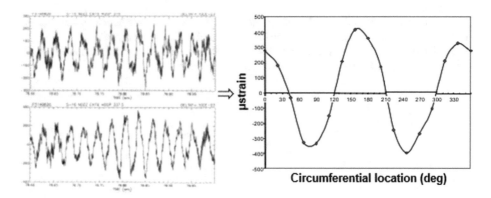

**Fig. 4.11**  Strain gage data from hot-fire test

transient 2ND mode [13]. It is clear that a thorough knowledge of the nozzle modal characteristics is critical to assessing the impact of either the transient or steady-state side load condition.

## References

1. NASA Image Galleries, https://www.flickr.com/photos/jurvetson/51431576323
2. Lepore F, Flow Induced Vibrations in the SSME Injector Heads, Rockwell Report 91–157, June 1991.
3. Goetz O, Monk J, Combustion Device Failures During Space Shuttle Main Engine Development, 5th International Symposium on Liquid Space Propulsion Long Life Combustion Devices Technology, October 27–30, 2003, Chattanooga, Tennessee.
4. Brown, A. M., Sullivan, R.M., Dynamic Modeling and Correlation of the X-34 Composite Rocket Nozzle, NASA/TP-1998-208531, July 1998.

5. Brown, A.M., Temperature Dependent Modal Test/Analysis Correlation of X-34 FASTRAC Composite Rocket Nozzle, Journal of Propulsion and Power, Vol. 18, Num. 2, March-April 2002.

6. M. Baker and C. Pray, Understanding Critical Dynamic Loads for Nozzle and Nozzle Extension Design, 47th AIAA/ASME/SAE/ASEE Joint Propulsion Conference & Exhibit, Feb. 2022.

7. J2-X Robust Nozzle Extension Design NAR, 17 Nov 2010, p. 111 not public.

8. Metallic Materials Properties Development and Standardization (MMPDS)-08, Battelle Memorial Institute, April 2013, 6–132.

9. Brown AM, Ruf J, Reed D, Keanini R, Characterization of Side Load Phenomena using Measurement of Fluid/Structure Interaction, 38'th AIAA Joint Propulsion Conference, 7–10 July 2002, Indianapolis, Indiana, AIAA paper #2002-3999.

10. Matsuyama K, Ito T, et al., H-IIA Rocket Engine Development, Mitsubishi Heavy Industries, Ltd. Technical Review Vol. 39, No. 2 (Jun. 2002).

11. Deck S, Nguyen AT, Unsteady Side Loads in a Thrust-Optimized Contour Nozzle at Hysteresis Regime.

12. Shi JJ, Rocket Engine Nozzle Side Load Transient Analysis Methodology—A Practical Approach, Structures, Structural Dynamics, and Materials Conference, 2005.

13. Blades EL, Baker M, Pray CL, Luke EA, Fluid-Structure Interaction Simulations of Rocket Engine Side Loads, 2012 SIMULIA Customer Conference.

# Rocket Engine System Loads

# 5

## 5.1 Introduction and Requirements

The calculation of the "loads" at interfaces throughout a LRE is one aspect of the SD of LRE's that is most similar to "loads" analysis throughout the rest of the launch vehicle. In general, dynamic forces are applied to various locations on low-fidelity FE models of sections of or the entire engine, and resolved total forces (called "loads") at interfaces are then given to a detailed stress analysis team; they in turn use them in a high fidelity static model of the structure to determine stress, strain, and HCF margins to the allowable capability, applying various safety factors as required. There are two main difference between LRE and launch vehicle loads. First, the frequency range of interest is an order of magnitude higher than launch vehicles (from 0 to 2000 Hz, compared to 0–200 Hz). Second, there are very discrete sources of random and harmonic excitation within a LRE, while the excitations on a launch vehicle are generally wide or narrow-band random only and are either mechanically transmitted through the structure from sources in the vehicle (LRE's and Solid Rocket Boosters), or externally applied due to the complex dynamic aero-acoustic field around the vehicle.

An additional source of excitation on LRE's is externally applied from the launch vehicle through the gimbal and propellant feed lines. This is particularly an issue for smaller upper-stage engines; they have loading transmitted during the boost phase while they are not operating, and even during operation they may be small enough that internally-generated engine excitations and resulting loads do not overwhelm those coming from the vehicle, unlike large boost engines.

The first step in the analysis is to understand the structural and performance requirements on the engine. The specified environments at the engine/vehicle interfaces (the gimbal, feedlines, and thrust control actuators), required engine mass properties, and maximum design conditions (MDC) will be identified in the "Interface Control Document"

© The Author(s), under exclusive license to Springer Nature Switzerland AG 2022    131
A. M. Brown, *Structural Dynamics of Liquid Rocket Engines*, Synthesis Lectures
on Mechanical Engineering, https://doi.org/10.1007/978-3-031-18207-5_5

(ICD). The structural margin requirements should be identified in a "Structural Analysis Criteria" (or "Plan"). One of these documents will also identify critical mission parameters such as mission and life cycle requirements and definition of various program design philosophies, such as whether a two or three-sigma value for random loads should be used in the program for various applications.

## 5.2    Engine System Loads Model

After requirements definition, the propulsion systems group will generate a very rough concept of the engine, including the turbopumps, thrust chamber assembly, nozzle, gas generator or pre-burners, thrust vector control actuators, gimbal, and ducting. The engines systems loads process can then be initiated, and the first step is the creation of an engine system FEM. An example of a finite element model built specifically for engine system loads generation is the Fastrac engine model shown in Fig. 5.1 [1]. For first-cut system models, and even sometimes beyond, all the system components (even turbopumps) are built using beam elements. This simplification reduces complexity while still capturing the essential bending and torsional motion that drive the resolved forces at the interfaces. The inertial properties of the beam elements are generated using high fidelity models of the specific components, or hand calculations of the simpler ones, such as ducts with constant cross sections.

There can be complications in the system model. As discussed in Chap. 4, the nozzle in the Fastrac was almost entirely a composite, and the Young's modulus changed significantly with temperature. This in turn affects the nozzle modes and the engine loads since the nozzle is the structural backbone of the engine. For the final set of design loads, a model was created for each 10 s increment of operation to reflect these changes [2].

A similar situation arises for the baseline RL-10, which, as mentioned previously, has the nozzle extension stowed during first stage ascent and then deployed with screws for upper stage operation. The boost stage ascent vehicle induced loads on the engine are not trivial, so a model of the stowed configuration must be created for analysis during that mission phase.

The initial J2-X dynamic loads model is shown in Fig. 5.2a, and a 3-D cross-sectional representation of a similar beam model of the RS-84 is shown in Fig. 5.2b. They are composed mostly of beam, plate, and spring elements, with some plate and solid elements [3]. The nozzle nodal diameter modes are generally not important for the vehicle dynamics, so it is modeled entirely with beam elements which only will capture its bending modes. Beam elements can use ring cross-sections for the inertia properties, so the 3-D visualization of these cross sections shows that the model reflects the actual geometry reasonably well.

Boundary conditions are always tricky for the assembly, but there are a number of methods to account for this difficulty, as discussed in Sect. 2.8. Depending on the amount

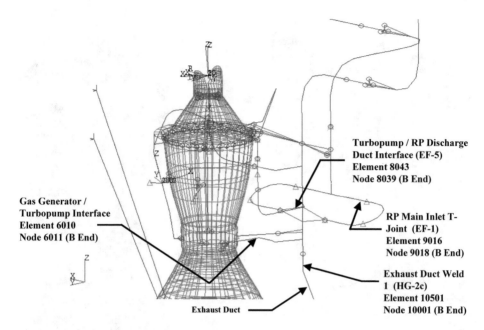

Turbopump / RP Discharge
Duct Interface (EF-5)
Element 8043
Node 8039 (B End)

Gas Generator /
Turbopump Interface
Element 6010
Node 6011 (B End)

RP Main Inlet T-
Joint  (EF-1)
Element 9016
Node 9018 (B End)

Exhaust Duct Weld
1  (HG-2c)
Element 10501
Node 10001 (B End)

Exhaust Duct

**Fig. 5.1**  Fastrac engine system loads model

of fidelity required, models of the hot-fire test facility "stiff-arm" (the mounting brackets), the vehicle thrust structure models, and the facility line may all be necessary.

The system loads model should be verified by modal test, to be discussed in Chap. 7, as soon as the engine is fabricated. The testing can be divided into modal surveys for the turbopump housing, the nozzle, and the entire engine assembly. These surveys will be used to produce the correlated engine loads model, which will be particularly important for verification of the internal engine modes and damping as well as interface boundary constraints and damping both in the vehicle configuration and in the test stand configuration.

## 5.3     Loads Events and Types

As with launch vehicles, the loads on an engine do not occur just during operation, and the loads during operation are not constant. Although the operational loads generally are the design driver, peculiar combinations of forcing functions and frequency can even make transportation loads (carrying the engine from the manufacturer to the launch vehicle for integration via truck and barge) the design driver for certain components. A list of possible loads phases is shown in Table 5.1 [4].

The forcing functions themselves, both static and dynamic, can also be categorized and tied to particular events from Table 5.1. We will discuss only the dynamic components in

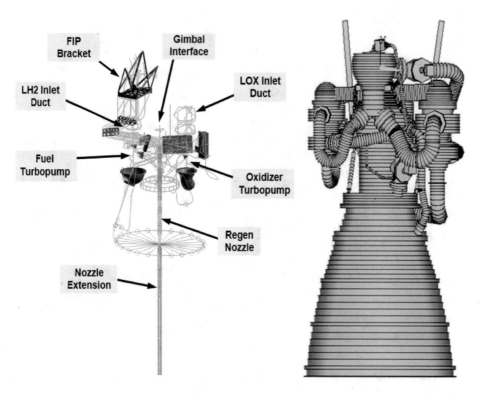

**Fig. 5.2**  J2-X engine system loads model for launch vehicle integration; **a** beams in 1-D visualization; **b** beams with 3-D visualization

| **Table 5.1**  Typical loads events for upper stage engine | |
|---|---|
| | Transportation |
| | Other pre-launch (Rollout, etc.) |
| | Booster liftoff |
| | Max Q (Dynamic Pressure) |
| | SRB separation |
| | Booster separation |
| | Altitude startup |
| | Main stage operation |
| | Shutdown |
| | Orbital or re-entry restart |

**Table 5.2** Engine forcing functions

| Vehicle induced |
| --- |
| Booster random (mechanical, acoustic, aerodynamic) |
| Booster harmonic (mechanical, acoustic, aerodynamic) |
| Separation shock |
| Pogo |
| Engine induced |
| Random |
| Harmonic |
| Shock |
| Nozzle transient sideloads |

this text, but not including shock loads or pogo, which are each entire disciplines. Some of these loads are shown in Table 5.2.

## 5.4 Characterization of Environments

Like the seemingly endless launch vehicle loads-cycle analyses, calculation of the engine system loads is also an iterative procedure. The determination of appropriate boundary conditions of the thrust vector control actuators, gimbal, and ducting, which are the interfaces with the launch vehicle, are critical. If the response to forces transmitted from the vehicle are being evaluated, those locations will be fixed-base excitation spots, but if response to engine-source excitations are being evaluated, a more compliant interface using spring elements or a dynamic-interface modal multi-point-constraint methods discussed in Sect. 2.8 could be used.

The actual forces acting on the engine, especially the self-induced ones, are very difficult to quantify. The random forces resulting from combustion processes of the thrust chamber and gas generator/fuel pre-burners especially are beyond the capabilities of physics-based CFD. The frequencies of the harmonic forces resulting from unbalances in the turbomachinery can be predicted, but physics-based calculation of the magnitudes of those loads, which are mechanically transmitted through the bearings and turbopump housings, are not tractable.

We instead use acceleration response measurements to back-calculate the steady-state forces used for excitation of the engine system model. The acceleration frequency domain PSD response for a typical test is shown in Fig. 5.3. The spikes are the sinusoids immersed within the overall random spectra. Amplitudes for these spikes are obtained by taking the area underneath it (the mean square, see Sect. 2.6), which can be converted into the peak sine response $G_{peak\text{-}sine}$, where $bw$ is the bandwidth of the measurement and $A$ is the amplitude of the measurement spike.

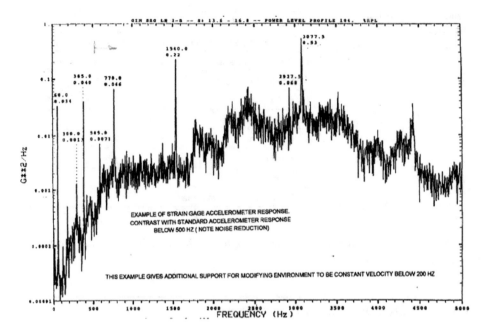

**Fig. 5.3**  Typical accelerometer response from engine hot-fire

$$G_{peak-\sin e} = \sqrt{2}\sqrt{bw * A} \qquad (5.1)$$

These loads are defined for both sinusoidal and random dominant cases for multiple zones around an engine, which are then documented in an "Environmental Criteria" document. This criteria is initially established using the Barrett Criteria (see next section), then is updated as the engine is tested and operated using measured hot-fire data from both strain-gages and accelerometers; the SSME criteria underwent 7 revisions during the Space Shuttle program! For each zone in the engine, the SSME criteria specifies sinusoidal amplitudes at specific frequencies in different directions, and random spectra PSD up to 2000 Hz; this upper limit is somewhat a relic of measurement capability in the 1980's, but it is generally a realistic bound of important system natural frequencies contributing to response. If a component has a natural frequency above 2000 Hz, recent test data should be used to evaluate it in lieu of the documented environment. An example comparing the SSME Block I Revision 5 and 6 criterion is shown in Fig. 5.4a for the random spectra and Fig. 5.4b for the sinusoidal.

As mentioned in Sect. 2.6, due to the nondeterministic nature of the random loads, the spectra must be "enveloped" to conservatively account for the peaks in the response. The straight lines in the zonal environment reflect the enveloping, which adds a significant amount of energy to the excitation, though, rendering the results sometimes

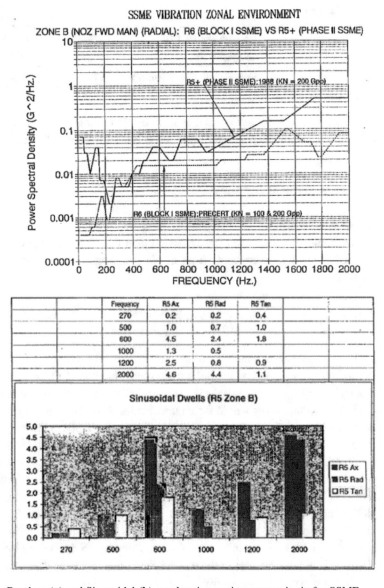

**SSME VIBRATION ZONAL ENVIRONMENT**

ZONE B (NOZ FWD MAN) (RADIAL): R6 (BLOCK I SSME) VS R5+ (PHASE II SSME)

| Frequency | R5 Ax | R5 Rad | R5 Tan |
|---|---|---|---|
| 270 | 0.2 | 0.2 | 0.4 |
| 500 | 1.0 | 0.7 | 1.0 |
| 600 | 4.5 | 2.4 | 1.8 |
| 1000 | 1.3 | 0.5 | |
| 1200 | 2.5 | 0.8 | 0.9 |
| 2000 | 4.6 | 4.4 | 1.1 |

**Sinusoidal Dwells (R5 Zone B)**

**Fig. 5.4** Random (**a**) and Sinusoidal (**b**) acceleration environment criteria for SSME

extremely over-conservative. Numerous efforts have been pursued, therefore, to reduce the conservatism of engine system loads.

In addition, responses clearly are not the same thing as the forces that we would like to have, but they can be used as base-drive excitations in the techniques discussed in Chaps. 2 and 4. Since they are not forces, but correlated and dependent responses, though, several

**Table 5.3** F1 and RS-84
barrett parameter comparison

|  | F1 | RS-84 |
|---|---|---|
| Vacuum Thrust T (lb) | 1,748,400 | 1,150,636 |
| Weight W (lb) | 17,000 | 14,500 |
| Exhaust velocity V (ft/sec) | 9730 | 10,862 |

problems arise, which I will discuss in coming sections. The first question though, is how do you get these measurements from an engine when it hasn't been built yet?

### 5.4.1  Barrett's Criteria for New Designs

For a new design, a frequently-used method to generate an initial estimate of both the random and harmonic components of the excitation uses the "Barrett Criteria" [5]. This method uses acceleration measurements from a reference engine $r$ for those loads and scales them using

$$G_n(f)\frac{g^2}{hz} = G_r(f)\sqrt{\frac{T_n V_n W_r}{T_r V_r W_n}} \qquad (5.2)$$

where $G_n(f)$ is the acceleration amplitude of the new engine $n$ as a function of the amplitude of the similar engine scaled by the turbopump speed, the nozzle thrust $T$, the nozzle exhaust velocity $V$, and the engine weight $W$. The original equation also has terms for the weight of individual components within the engine, but those are generally neglected. An example is shown below, using parameters from the Saturn-V F1 first stage engine as the reference engine for a first-cut dynamic environment for the RS-84 Lox Kerosene engine of the early 2000's. The RS-84 design Isp was 265.2 s, so since the Isp of the F-1 was 306 s, and both engines use Kerosene as their fuel, they are classified to be in the same family. Table 5.3 shows the parameters of the two engines.

The result of applying the Barrett Criteria is $G_{RS84} = 0.93 \ G_{F1}$, indicating that the loads should be fairly similar between the two engines. The criteria was also used to create the first set of J2-X environments by scaling with the RS-68, an engine used in the Delta IV launch vehicle. For use in the random environment, the amplitude should be squared to obtain a power spectral density (discussed in Sect. 2.6).

### 5.4.2  Use of Measured Data for Existing Designs

There are several methods for using the accelerations as excitations. An extensive evaluation needs to be undertaken for every new engine on which of these is appropriate based on an understanding of the assumptions, approximations, and limitations of the techniques. The first methodology is to create component-only models and to base-drive those components from "fixed" interface locations; this is the same method as used for the analysis of combustion devices described in Chap. 4. In this case the components would be a single duct, and the opposite bases, which might be the fuel tank and the turbopump inlet, would both be fixed and undergo the same base drive using excitation frequencies and magnitudes from the closest excitation source.

For the random loads, the Miles Equation (discussed in Sect. 2.6), can be used instead of the base drive methods available in a FE code, if the mode of interest is well separated and if the frequency spectrum of the excitation region around the mode is close to being constant (white noise). Alternatively, the RMS of the random load across the entire spectra can be applied as a static acceleration force, which assumes essentially non-resonant conditions. A set of NASA recommendations from 2001 [6] specifies using Miles equation or base-drive random FE analysis for ducts and associated hardware, including brackets, stiffeners, and actuators; and the equivalent static acceleration for interfaces of combustion devices, which generate random forces themselves. The random load for interfaces between the engine system and the main combustion chamber, and at the gimbal/vehicle interface are considered negligible. For these analyses, the criteria will specify a specific damping value; 3.3% damping was used for the SSME (and now the RS25) for the first 6 modes, and 5% for higher modes. Since the time when these recommendations were written, random analysis methods in finite element codes have improved to the point where multi-point inverse force excitation methods now predominate, but these simpler methods are more appropriate in the early stages of the design.

For the sine components of the loading, a 10% frequency band about each discrete frequency (as shown in Fig. 5.4) is assumed to account for uncertainty in either the excitation or natural frequencies. Also, the forcing functions should be applied one axis at a time, but the results for each excitation, which will have components in each axis, should be summed on an axis-by-axis basis.

The component methods are clearly inaccurate as they ignore the dynamic interaction within the engine propagating through the supposedly "fixed" bases, and the multiple loads sources. However, in practice they provide somewhat reasonable results when compared with hot-fire strain gage data (i.e., conservative but non-overly-conservative), and in fact have been shown to be less overly-conservative than analysis using a multi-base-drive of a system model (discussed below) in some cases (the SSME and the Fastrac) [7].

In an attempt to provide an alternative to the overly conservative results of either the component model or engine system model base acceleration techniques, particularly for engines with multiple excitation sources of random loads, methods have been developed to calculate forces that would reproduce the measured accelerations. A good bit of the development of random analysis of structures with multiple sources arose from the earthquake engineering field [8], and until recently the best reference textbook in this field of LRE system loads analysis was "Dynamics of Structures and Earthquake Engineering" by Chopra [9]. Let's first look at the multiple-base enforced acceleration excitation method.

Summarizing the development by Christensen and Brown, which is similar to the method described in Sect. 2.5.8 but now extended to multiple "base" sources, the engine DOF's can be partitioned into DOF's $e$ that are enforced (the sources) and free DOF's $f$, much like Craig-Bampton partitioning discussed in Chap. 2.

$$\{x(t)\} = \left\{ \begin{array}{c} \{x_f\} \\ \{x_e\} \end{array} \right\}$$

which when substituted into the equations of motion along with a zero external force, yields

$$[M_{ff}]\{\ddot{x}_f\} + [C_{ff}]\{\dot{x}_f\} + [K_{ff}]\{x_f\}$$
$$= -[M_{fe}]\{\ddot{x}_e\} - [C_{fe}]\{\dot{x}_e\} - [K_{fe}]\{x_e\}$$

where the right-hand side of the equation, which can be calculated directly from the specified enforced acceleration, equals the equivalent force required to enforce that acceleration $\{F_{eq}\}$. We can now obtain the PSD (using the standard $S_{xx}(f)$ terminology here) of the non-specified DOF's using the standard spectral density relationship with the PSD of $F_{eq}$

$$[S_{xx}(f)] = [H^*(f)][S_{F_{eq}F_{eq}}(f)][H(f)]^T$$

where the $H^*$ denotes the complex conjugate. The response is composed of a "pseudo-static component" $\{x_{fs}(t)\}$, which is due to the stiffness-only reaction, i.e., the inertial effect removed as if the motion was proceeding extremely slowly, and a dynamic part $\{x_{fd}(t)\}$ which is the inertial component only. As shown in Sect. 2.5.8,

$$\{x_{fs}\} = -[K_{ff}]^{-1}[K_{fs}]\{x_s\}$$

so the PSD calculation is

$$[S_{Xfs}(f)] = \frac{1}{(2\pi f)^4}[K_I][S_{aa}(f)][K_I]^T \tag{5.3}$$

where

$$[K_I] = -[K_{ff}]^{-1}[K_{fs}]$$

and $[S_{aa}(f)]$ = acceleration PSD's at source locations. The PSD response of the dynamic portion, $[S_{xfd}(f)]$ can be obtained similarly.

The pseudo-static component can be significantly over-estimated at low frequencies. There are two reasons for this; first, the displacement equals the acceleration integrated twice, so as the frequency gets smaller, any error in the measurement, which is generally non-trivial for most accelerometers at low frequencies, become magnified. This is seen clearly in Eq. 5.3 where the frequency in the denominator is raised to the 4th power. More importantly, when there are multiple sources of excitation relatively close together in an engine, for example in a monolithic powerhead where the turbopump (or pumps) are very close to the thrust chamber, the duct in-between is excited by each of these assumed-to-be completely uncorrelated, independent sources. At low frequencies (i.e. as the response becomes more static), the quasi-static portion will start dominating the response, and the above error yields a much larger displacement between the two sources than actually exists, producing a huge stress. In reality, the sources are correlated due to the high stiffness of the structure, the relative displacement is actually very small, and the stresses are therefore much smaller as well.

The "equivalent applied force methods" offer a hoped-for improvement in the drawbacks of the direct enforced acceleration methods. In these, the goal is to generate a set of uncorrelated force PSD's that replicates the specified acceleration field. The first step is to calculate a transfer function matrix $[T(f)]$ relating spectral densities of known acceleration PSD's (subscripted aa) to spectral densities of unknown forces at those locations (subscripted FF)

$$\{S_{aa}(f)\} = [T(f)]\{S_{FF}(f)\}.$$

$[T(f)]$ can be calculated by individually applying a unit force PSD onto the model, and the resulting spectral acceleration response is equal to that appropriate value on the diagonal of $[T(f)]$. To obtain $\{S_{FF}(f)\}$, we unfortunately cannot simply invert $[T(f)]$ because the result may not be positive definite, so instead a simple division of each element in $\{S_{aa}(f)\}$ by the diagonal elements of $[T(f)]$ is performed. This step is where approximations enter as it assumes the accelerations are applied one at a time and one direction at a time. Some more advanced methods have been developed recently that use analytically-calculated cross-spectral densities relating measured locations and all the potential sources to supplement $[T(f)]$ and remove these approximations; this method was applied in the loads analysis of the Space Launch System Exploration Upper Stage [10].

Christensen presents data obtained from a test using an excess Fastrac engine and its model as a testbed, with known inputs from independent shakers on each orthogonal axis and measured strain and acceleration responses (see Fig. 5.5). The artifact of the excessive pseudo-static contribution is clearly seen in high response at low frequencies in both the enforced acceleration and equivalent applied force methods. In the enforced acceleration method, though, this component can be removed, leaving only the dynamic portion, which significantly improves its results as shown in Fig. 5.6, where the green curve for that technique shows a much better match with the dark blue test and red forced response using the known input than the light blue equivalent force. Using only the dynamic portion of the enforced acceleration method for loads analysis using a complete engine model with multiple excitation sources is therefore recommended.

**Fig. 5.5** Fastrac engine system loads testbed. The 3 shakers can be seen in grey

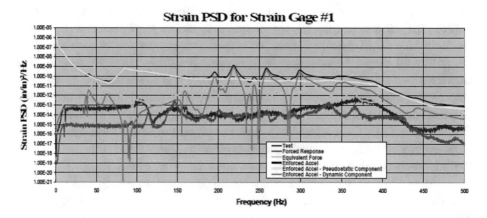

**Fig. 5.6** Strain gage results from FASTRAC testbed of engine system loads methods

## 5.5    Engine System Loads Tables and Combination of Random and Harmonic Loading

The purpose of the complicated analysis presented in Sect. 5.4 is to provide a relatively simple output, a table for each interface component showing the resolved bending, axial, torsional, and shear forces resulting from random, harmonic, and static loading in the X, Y, and Z directions. In some cases, the directional forces are resolved into a single vector direction. A typical loads table from the FASTRAC is shown in Table 5.4. These loads are used to determine the stress margin of these joints, brackets, and other interfaces using high fidelity finite element static analysis. This information is collated in a "Structural Loads and Environments" document and distributed to the engine team. A flowchart of a complete engine system loads process is shown in Fig. 5.7.

An obvious question when using this table is how are loads from the harmonic analysis and those from the random analysis combined to calculate the margin. A detailed examination of this process was undertaken by Brown and McGhee [11]. There are several simple closed form equations that perform this calculation, the main ones being the "standard method", in which the design load $DL$ equals the sum the amplitude of the sine load $A_{sin}$ and 3 times the RMS of the random load. For a random load that is normally distributed (the general assumption), the RMS equals one standard deviation $\sigma_{ran}$, so

$$DL = A_{\sin} + 3\sigma_{ran}.$$

The inherent assumption in this equation is that the peak of the harmonic response occurs simultaneously with the peak random value, which as previously mentioned is

**Table 5.4**  Loads table for two FASTRAC components

| Glue bracket 3 | Shear 1 | Shear 2 | Axial | Bending 1 | Bending 2 | Torque |
|---|---|---|---|---|---|---|
| GB-3 | (lbs) | (lbs) | (lbs) | (in-lbs) | (in-lbs) | (in-lbs) |
| Sine X | 97.1 | 7 | 0 | 3 | 77.7 | 71.7 |
| Sine Y | 90.6 | 7 | 0 | 3 | 98.1 | 70.2 |
| Sine Z | 119.2 | 5.1 | 0 | 0.2 | 78 | 52.3 |
| Sine peak (RSS) | **178.4** | **11.1** | **0** | **4.8** | **147.5** | **113.1** |
| 3 sig random X | 450.5 | 112.8 | 0 | 16.4 | 25.2 | 1475.5 |
| 3 sig random Y | 781.2 | 65.5 | 0 | 8.8 | 41 | 828.5 |
| 3 sig random Z | 155 | 0.6 | 0 | 3 | 1100.6 | 5.7 |
| Random Peak (RSS) | **915** | **130.4** | **0** | **19** | **1101.7** | **1692.1** |
| Bracket 3 (lower support) | | | | | | |
| SB-6 | | | | | | |
| Sine X | 18 | 7.6 | 10.7 | 8.2 | 17.1 | 2.2 |
| Sine Y | 12.3 | 4.1 | 10.3 | 7.3 | 11.2 | 1.3 |
| Sine Z | 10.9 | 12.3 | 7.7 | 2.9 | 27.8 | 2 |
| Sine peak (RSS) | **24.4** | **15.1** | **16.7** | **11.4** | **34.5** | **3.9** |
| 3 sig random X | 34.7 | 333.3 | 5.7 | 85.1 | 1348.8 | 51.7 |
| 3 sig random Y | 59.9 | 191.5 | 10.1 | 144.9 | 774.9 | 29 |
| 3 sig random Z | 12 | 1.3 | 11.3 | 83.2 | 6.3 | 0 |
| Random peak (RSS) | **70.2** | **384.4** | **16.2** | **187.5** | **1555.6** | **59.2** |

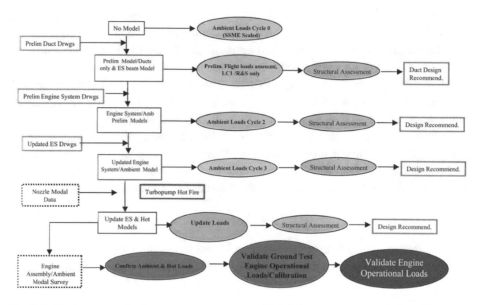

**Fig. 5.7**   Engine system loads process

considered to be 3*sigma. This assumption is always conservative, but can be grossly so, leading to an erroneous conclusion that a component doesn't pass requirements.

The second simple closed form technique frequently used is the "3*ssMS method" in which

$$DL = 3\sqrt{(\sigma_{\sin})^2 + (\sigma_{ran})^2} = 3\sqrt{\frac{A_{\sin}^2}{2} + (\sigma_{ran})^2}$$

since the standard deviation of the sine is the amplitude divided by $\sqrt{2}$. This method in effect uses the sine as if it were a normally distributed random variable, which has a much wider distribution than is actually the case, and so also is always conservative. The third of these formulas, called the "Peak Method", was proposed by Steinburg [12] for use in electronics components under combined loading, and is

$$DL = \sqrt{(A_{\sin})^2 + (3\sigma_{ran})^2}.$$

Since the random portion of these loads is substantial, we would like to have a probability associated with the design load, which none of these methods achieve. With the goal of obtaining a mathematically accurate method for combining these loads into a $DL$ value that can be tied to a probability level, a method using the probability density functions (PDF) of both the sine and random components was therefore investigated. Probability theory shows that convolution can be applied to these two PDF's resulting in a joint PDF of the design load

$$\phi(DL) = \int\limits_{-A}^{A} \left( \frac{1}{\sigma_{ran}\sqrt{2\pi}} \int\limits_{-\infty}^{DL-y} \exp\left( \frac{-\left( x/\sigma_{ran} \right)^2}{2} \right) dx \right) \frac{1}{\pi A \sqrt{1 - \left(\frac{y}{A}\right)^2}} dy$$

This equation is not analytically integrable, but the symbolic mathematics code Mathematica$^{©}$ can numerically solve it, and solutions for a number of combinations of random and sine amplitudes were obtained. Since this product of a deterministic function with a random variable is also a random variable, the specific targeted level sought was 99.865%, which is the three-sigma level for a normal distribution (called here the "Equivalent 3-Sigma" value). For verification, a Monte Carlo analysis for the 3-sigma level for *DL* was also computed and showed agreement with the convolution integral with less than 0.1% difference. The results show that compared to this result, which is taken to be the most accurate solution, the "standard" method overpredicts the *DL* by between 6 and 14%, while the "3*ssMs" method overpredicts by between 0 and 62%. On the other hand, the "Peak" method consistently underpredicted, ranging from $-1\%$ to over $-18\%$, so is significantly under-conservative, never a desirable result.

Upon further examination of the amount of overprediction of the "3*ssMS" method, the plotted solution in terms of the ratio of the random and sines shows a consistent curve for the specific probability level. This curve is therefore fit with a polynomial which can be much more easily implemented and solved than the original symbolic equation. For the 99.865% value, the overprediction is

$$overshoot = \exp\left(-1.53546/x\right) * \left( \begin{array}{l} 4.60691 * 10^{-11}x^7 - 1.193679 * 10^{-8}x^6 + \\ 1.19197 * 10^{-6}x^5 - 6.04117 * 10^{-5}x^4 + \\ 1.69965 * 10^{-3}x^3 - 0.0271317x^2 + 0.241154x \end{array} \right)$$

(5.4)

which is used in the following equation for the DL, where $x$ is ratio of the random standard deviation to the sine amplitude.

$$DL = \frac{3\sqrt{\frac{A_{sin}^2}{2} + \sigma_{ran}^2}}{1 + overshoot}.$$

(5.5)

When compared against the original analytical expression, this curve fit yields errors of less than 0.6%. A similar calculation can be made for other desired probabilities, e.g., 97.5%, which is the 2*sigma level frequently used in HCF evaluation.

There have been several proposals in the literature to represent the sine response within the random spectra and to just perform a single random analysis and loads calculation [13]. This leads to yet another instance of excessive conservatism, though, so it is recommended keeping them separate and combining the results as described above.

## 5.6     Lessons Learned

There are a number of general "lessons learned" during engine system loads analysis at NASA/MSFC included in the 2001 recommendations previously referenced which may be helpful

- steady-state, not oscillatory, loads tend to dominate for most brackets and interfaces, but external interface loads are dominated by quasi-static, such as start and shutdown sideloads transients.
- A design coupling the turbopump with the nozzle below the thrust chamber throat will frequently generate substantial loads and design sensitivity.
- No single load case will yield the maximum loads for every interface.
- Vehicle acceleration loads are usually enveloped by engine system self-induced loads
- The influence of the nozzle on the engine system loads is significant, so correlation with modal test is critical, particularly if the nozzle design is outside of the experience base (such as composite or channel-wall nozzles)
- Adequate statistical hot fire data characterization is key to performing test versus analysis verification, e.g., comparing a test hot-fire RMS with an analytical RMS.

## 5.7     Nozzle Sideloads During Engine Startup and Shutdown

The nozzle sideloads phenomena was discussed in Sect. 4.4 relative to how it affected the nozzle capability itself. Even more critical, though, is the reacted load of the transient sideload condition in the thrust vector control (actuators) during startup and shutdown. For example, the first estimate of this load for the J-2X engine was close to a million pounds, a huge value that would have resulted in a massive re-design. While this value was reduced significantly, it's an indicator of how large the load is. At this time, the conservative skew-plane method is still used for the initial estimate of this quasi-static force, which is placed at worst-case circumferential locations in the nozzle section of the engine system loads model. Revisions incorporating comparisons to sideload measurements of previous engine programs are then scaled to generate more refined values in later loads cycles. These measurements generally are taken with a force transducer placed at the gimbal in the engine test cells; e.g. a very large database was collected for SSME that can be used for comparison. A more accurate analytical estimate would still be beneficial, though, as this sort of empirically-based engineering solution always has limitations.

## 5.8    Excessive Turbomachinery Response in Orbital 3 Launch Failure

In addition to calculating loads for design, a detailed loads model can be invaluable for failure investigations, as in the Orbital 3 launch failure of the Orbital ATK Antares Launch vehicle in October 2014 [14]. The launch was part of the new commercial cargo program taking cargo to the International Space Station. Fifteen seconds into the flight, the main propulsion system (engine and propellant feedlines) exploded (see Fig. 5.8). There were no injuries, but there was extensive damage to the launch pad and the entire launch vehicle and payload was destroyed.

NASA/MSFC was called in to help determine the cause of the failure by examining the engine, which was a slightly-refurbished Russian NK-33 renamed the Aerojet AJ26. This engine had been adopted for use based almost entirely on a successful flight history (at least that is what Aerojet was told) but there was very little detailed understanding of the engine design. There had been a limited amount of data taken during flight, and one measurement showed very high acceleration levels at a fuel propellant line near the tank. It was hypothesized that the turbomachinery had dangerously high vibration levels as well, and that these accelerations were dynamically transmitted through the engine system to the fuel duct line where the measurement was taken. Since the existing engine system

**Fig. 5.8**  Orbital-3 antares launch vehicle failure 2014

**Fig. 5.9** Engine system loads
model coupled with antares

**Fig. 5.9** Engine system loads
model coupled with antares

model was extremely simplified, MSFC was asked to build a more hi-fidelity model, and
in addition, to incorporate a model of the fuel tank to determine a more accurate transmissibility, i.e., a turbopump excitation environment would be applied and the response
compared with the measurement at the fuel line to see if an excessive environment is
required to obtain a good match. Both a very detailed FEM and a somewhat simplified
FEM focusing only on the frequency range and locations of interest were built (simplified
model shown in Fig. 5.9).

The resulting transfer function was approximately 0.7 g response for a 1 g loading at
the synchronous (1 N) of the turbopump. This value was not only consistent with the fuel
line response measured during the abbreviated flight, but also more detailed measurements
taken during ground-testing (Fig. 5.10). The hypothesis of large turbomachine vibration
was proven consistent with the available measurements, therefore, and along with other
indicators used to conclude that it was the cause of the failure.

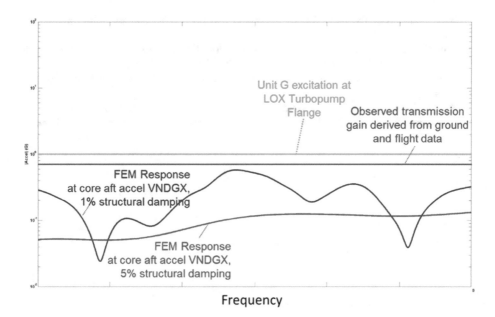

**Fig. 5.10** Comparison of frequency response measured versus model

## References

1. Christensen ER, Frady G, Mims K, Brown AM, Structural Dynamic Analysis of the X-34 Rocket Engine, AIAA-98-2012.
2. Frady G, Jennings JM, Mims K, Brunty J, Christensen ER, Engine System Loads Analysis Compared to Hot-Fire Data, 43rd AIAA Structures, Structural Dynamics, and Materials Conference, 22–25 April 2002, Denver, CO, AIAA-2002-1510.
3. Bishop J, Brown A., Ares I Upper Stage Engine Dynamic Loads Model, Feb. 24, 2010, NASA/MSFC Memo # ER41(10-005).
4. Passerini RM, J-2X Engine Loads and Environments Document—R1.0 Load Set, August 2006.
5. Barret, RE, Techniques for Predicting localized Vibratory Environments of Rocket Vehicles, NASA TN-D-1836, 1963.
6. Mims K, Brown AM, Rocket Engine System Structural Dynamics Loads Development Guidelines, Unpublished MSFC Presentation, August 29, 2001.
7. Christensen ER, Brown AM, Frady GP, Calculation of Dynamic Loads Due to Random Vibration Environments in Rocket Engine Systems, 48'th AIAA Structures, Structural Dynamics, and Materials Conference, Honolulu, Hawaii, April 23–26, 2007.
8. Leger P, Ide IM, Paultre P., Multiple-Support Seismic Analysis of Large Structures, Computers & Structures, Vol. 36, No. 6, pp. 1153–1158, 1990.
9. Chopra, AK, Dynamics of Structures, Theory and Applications to Earthquake Engineering, Prentice-Hall, New Jersey, 2001.
10. Baker M, Ignatius C, et. al, Process for Efficient Calculation of Rocket Engine Interface Random Vibration Loads Using Acceleration Time History Measurements from Hot Fire Test,

ATA Engineering Technical Paper - Spacecraft and Launch Vehicle Dynamic Environments Workshop, June 21–23, 2021

11. Brown, A. M., McGhee, D.S., Statistical Evaluation and Improvement of Methods for Combining Random and Harmonic Loads, NASA /TP-2003–212257, February 2003.

12. Steinberg, DS, Vibration Analysis for Electronic Equipment, John Wiley & Sons, New York, p. 249, 1988.

13. Shirr R, Yuen M, Shen L, 12–3–2015-TIM_loads_status.pptx, Aerojet Rocketdyne, 2015.

14. NASA Independent Review Team: Orb-3 Accident Investigation Report, Executive Summary, NASA 9 October 2015.

# Structural Dynamics of System Hardware and Propellant Feedlines

# 6

## 6.1  Introduction

This chapter discusses the structural dynamic evaluation of specific components in the engine system. Essentially, this covers anything not in the turbomachinery and combustion devices and includes the propellent feedlines all the way to the fuel tanks. It is quite similar in some ways to the previous chapter, especially the component level engine system loads analysis. However, here we discuss some specific components with unusual forcing functions or structural dynamic characteristics.

## 6.2  Components Upstream of Inducer Cavitation

As discussed in the introduction, the purpose of the pump side of the turbomachinery is to raise the propellant pressure from the relatively low pressure in the tanks to pressures high enough for combustion. This increase is done in a series of steps, the first of which is passage through an inducer. As discussed in Sect. 3.8.6, one of the forcing functions resulting from inducer operation is Higher Order Surge Cavitation, which quite efficiently propagates upstream from the inducer. As the magnitude of this forcing function is non-trivial and is somewhat harmonic, possible resonant response of structures in the duct must therefore be evaluated.

An example of a detailed analysis performed using this excitation is in the work by Brown and Mulder [1]. Following that development, the acoustic environment of the cavitation should be based on a measurement and analytical program to gain an understanding of the hydrodynamic field. The acoustic field is not limited to the HOC surge and rotating components, as described previously, but also has modulations of these frequencies at HOC + 2N as well as more pure tones at 1N and 2N. The predominant excitation amongst these is identified from test results and has a "semi-narrow band" spectral character, with a frequency band approximately 10% of the nominal excitation frequency. The description of the pressure wave, which spins as it propagates upstream, is

$$p(\theta, x, t) = a_{m,n} \cos[m\theta - 2\pi f t]e^{-k_{m,n}x} \tag{6.1}$$

where $a$ is the amplitude of the excitation, $\theta$ is the circumferential angle, $x$ is the axial distance upstream of the inducer, $m$ is the number of nodal diameters, and $n$ is the number of nodal circles. $k$ is an attenuation parameter quantifying how much the magnitude of the wave dies off with increased distance from the source, and is expressed as

$$k = \frac{2\pi}{c}\sqrt{f_{m,n}^2 - f^2}$$

where $f_{m,n}$ is the acoustic natural frequency of the $m,n$'th acoustic mode, $f$ is the nominal excitation frequency, and $c$ is speed of sound in the fluid.

For the development of the RS-25, this flowfield was applied at resonance on a number of upstream components in the feedline. This loading is over-conservative as the excitation is not purely harmonic but instead is narrow-banded, but the resonant analysis is a good screen. The only component that did not pass this analysis was the "pre-valve" (Fig. 6.1).

A transient analysis using Eq. 6.1 as a scaling equation on measured pressure amplitudes from the inducer could be used to map out the complete temporal history of each point, and this would be the most accurate technique for assessing the true stress response of the pre-valve. As this type of analysis is close-to-intractable, though, a single DOF modal approximation of the structure was created, and the transient applied using an ordinary differential equation solver. These results are explained in the referenced paper and show the pre-valve to be acceptable.

A similar structure upstream of the inducer is a bellows "flowliner," used to protect the bellows itself from unsteady flow (see Fig. 6.2). In 2002, a crack in a flowliner in this duct was detected in the SSME, and since potential consequences of a complete flowliner failure could be catastrophic, the entire Space Shuttle fleet was put on hold until the root cause could be determined [2]. The resulting examination proved that several flow instabilities, including HOC, probably excited flowliner modes, which were quite numerous in the excitation frequency region (Fig. 6.3). One of these excitations was acoustic "Helmotz" resonance lock-in as the flow passed over the slots of the flowliner.

A series of dynamic tests were performed to validate the models of the flowliner, and using these, measured strains during hot-fire testing could be extrapolated to obtain a better understanding of the fatigue loading. This led to the development of a different non-destructive evaluation technique which could catch much smaller defects, so an enhanced safe service life and associated inspection intervals significantly reduced the risk of cracks.

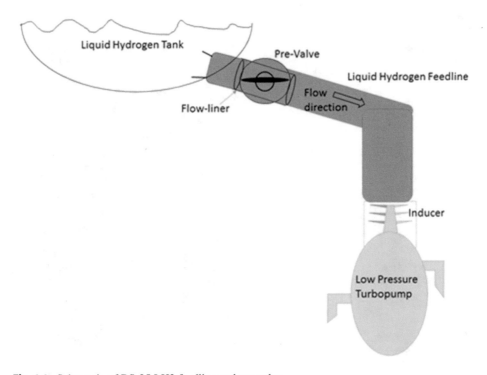

**Fig. 6.1**   Schematic of RS-25 LH2 feedline and pre-valve

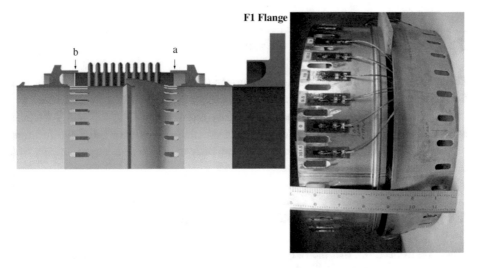

**Fig. 6.2**   Flowliner cross section in duct shielding bellows, and instrumented flowliner

**Fig. 6.3** One of numerous flowliner modes in unsteady flow and acoustic excitation range

## 6.3     Vortex Shedding of Probes

There are a number of cylindrical temperature and pressure probes in ducts in every LRE for validation of these critical parameters. As these probes are cantilevered in a liquid flow field, the vortex shedding phenomena previously discussed in Chaps. 3 and 4 becomes particularly problematic. The shedding frequency should be at least estimated for every probe and compared with the probe natural frequency to ensure an adequate margin from feedback lock-in (limit cycle response).

## 6.4     Fluid-Induced Vibration of Duct Bellows

I mention in Sect. 6.2 that the purpose of the shuttle flowliner is to shield the bellows to try to mitigate typical problems due to its own fluid-structural dynamic interaction, which is called "Fluid-Induced Vibration" (FIV). Bellows in rigid ducts are used to allow articulation of the engine during thrust-vector control movement, and to reduce misalignment

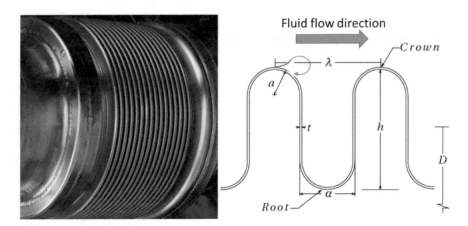

**Fig. 6.4** Bellows and cross-section of a convolute

loads. The cross section of a bellows is shown in Fig. 6.4, and the problematic location is at the "convolutes" [3].

Like vortex shedding, the flow field on the inside of the duct will shed vortex-like fluid disturbances as it passes over the convolutes. These flow fields can grow for certain values of the geometric parameters λ and α (shown in the figure) and interact with axial "slinky" modes of the convolute, causing significant structural response and cracking at the root. An additional complication is that accurate finite element modeling of the convolutes is non-trivial, particularly when the convolute is fabricated in two separate layers. There has been a good bit of research and testing performed on this topic, and a seminal NASA technical memo by Tygelski, Smyly, and Gerlach established empirically based approximate conservative solutions for design to avoid significant convolute response. Less overly conservative methods have been studied and are available for review as well [4]; a well-documented updated methodology has recently been published by NASA but has controlled distribution (reference: Dorney D, Eddleman DE, Motion Magnification for Gimbaled Bellows, NASA Engineering and Safety Center Technical Assessment Report TI-21-01647, February 3, 2022).

## References

1. Brown, A. M., Mulder Andrew D, Single Degree-of-Freedom Modeling of SLS Liquid Hydrogen Pre-Valve Flow Guide to Enable Rapid Transient Analysis, 58th AIAA/ASCE/AHS/ASC Structures, Structural Dynamics, and Materials Conference, AIAA SciTech (AIAA 2017-1130).
2. Frady G, The Role of Structural Dynamics and Testing in the Shuttle Flowliner Crack Investigation, 46th AIAA Structures, Structural Dynamics, and Materials Conference, 18–21 April 2005, Austin, Texas.

3. Tygielski PJ, Smyly HM, Gerlach CR, Bellows Flow-Induced Vibrations, NASA TM-82556, 1983.
4. Higgins SL, Davis RB, Brown AM, Reduced-Order Modeling of Flow-Induced Vibrations in Bellows Joints of Rocket Propulsion Systems, AIAA SciTech Forum, 4–8 January 2016, San Diego, CA.

# Structural Dynamics in Testing of Liquid Rocket Engines

7

## 7.1 Introduction

There are three main types of structural dynamic testing of liquid rocket engines: modal testing, hot-fire testing, qualification and acceptance testing. In addition, whirligig testing is a specialized but valuable test for turbine bladed-disks. Each of these types play a vital role in the verification of the structural capability of LRE hardware. In the initial days of rocket engine development, verification of structural capability was reliant on testing for every single piece of hardware. As analytical methods have become more reliable and high-fidelity, this reliance has decreased significantly, with a tremendous savings in cost. However, there is still and always will be cases where testing is required.

## 7.2 Modal Testing

As discussed in Chap. 2, this type of testing is focused on validation of the basic structural dynamic characteristics of a component obtained through analysis, its natural frequencies and mode shapes. It is used extensively in all areas of structural dynamic evaluation in LRE's, so it was necessary to give the theoretical background in the earlier chapter. Here are, a few examples of specific LRE applications of modal testing.

A "ping" test of a bladed-disk is shown in Fig. 7.1 [1]. This differs slightly from a complete modal test in that only the individual blade natural frequencies are measured, not the mode shape. Due to the mistuning phenomena, quantifying these natural frequencies is critical for generating a specific mistuned configuration as well as for eventually

**Supplementary Information** The online version contains supplementary material available at https://doi.org/10.1007/978-3-031-18207-5_7.

A. M. Brown, *Structural Dynamics of Liquid Rocket Engines*, Synthesis Lectures on Mechanical Engineering, https://doi.org/10.1007/978-3-031-18207-5_7

**Fig. 7.1**  Ping test of bladed disk

generating the frequency variation statistics with enough samples. Frequently, each blade of interest is isolated by either mass-loading the other blades with magnets or putting foam between the other blades; there is, however, some debate as to whether isolation is necessary for either modal identifiability or repeatability.

Another interesting modal test of a LRE component was on one of the SSME High Pressure Turbopump housings in the mid 90's. The model was critical for providing dynamic boundary conditions for the rotating shaft. One end of this housing consists of a very stiff flange where the pump is bolted to the rest of the engine. Modal analysis showed that a free-free configuration would give useful modes of the central portion of the housing, but the flange would behave as a rigid body. Since the flexibility of the flange is of importance in the overall response, the team determined that a large 1300 lb cylindrical "doughnut" shaped mass that happened to be on-hand could be bolted onto the flange, then the entire assembly would be placed on foam to simulate a free-free condition (Fig. 7.2, white insulation covers much of structure). The inertial of this huge mass, which can be easily modeled, allowed the flexibility of the flange region to be exercised in the mode shapes, as shown in Fig. 7.3.

On a much larger scale, modal testing of the entire LRE is also critical, particularly for accurate coupling of the model with the launch vehicle to enable accurate guidance, navigation, and control through the thrust vectoring system. An animated comparison of the analytical and tested mode shapes for the second nozzle nodal diameter mode from a modal test of the Space Shuttle Main Engine is shown in online resource 11.

An interesting modal test of a turbopump inducer is shown in Fig. 7.4. In this case, the inducer is submerged in a tub (actually a crawfish-boil pot!) of water to attempt to

**Fig. 7.2** SSME high pressure
turbopump housing with
doughnut

replicate the dynamic characteristics in liquid oxygen, which has an almost identical density. The extra complication of the liquid environment is necessary as structural/acoustic modeling is still in its infancy, and the effect of the liquid is substantial.

## 7.3 Hot-Fire Testing

A test of a rocket engine under full operating conditions, of course, is required to validate that the engine functions as intended and does not fail structurally. As a tool for structural dynamics, though, a hot-fire test (Fig. 7.5) is not the "proof is in the pudding" that many would believe it is. There are two main drawbacks to a hot-fire test for examining structural dynamics of LRE's. The first is that resonance is generally a low-probability occurrence, requiring a close alignment of an excitation frequency with a natural frequency, and both quantities are somewhat non-deterministic, especially the operating natural frequency. Ignoring structural dynamic effects, though, is fraught with peril. As mentioned in the introduction, both the SSME and RS68 engines logged a significant number of premature cut-offs due to resonance induced high cycle fatigue. The second reason is that a hot-fire test at sea level is not an accurate reproduction of the operating environment for a high-altitude engine's nozzle, so the side loads, discussed

**Fig. 7.3** Free-free 2nd bending mode at 548 Hz of SSME high pressure pump housing with "dough-nut" on right side

**Fig. 7.4** Modal testing of inducer in water

**Fig. 7.5**  RS-25 hot fire test

in Sect. 4.4, cannot be measured accurately unless a stand is built which incorporates a vacuum device to simulate operation at high altitude.

On the other hand, hot-fire tests do provide critical information for SD analysis and evaluation of LRE's. First, the environment used in engine system loads analysis is obtained by accelerometer and strain gage measurements taken throughout the engine, as discussed in Chap. 5. In addition, hot-fire test is critical for validating input parameters in a SD analysis, such as temperature, flowrate, and turbopump rotational speed.

Hot-fire tests also allow quantification of effects that may be too difficult to analyze, such as nondeterministic pits on turbine blades. A fully rigorous statistical evaluation of these effects would require an unreasonable amount of expensive hot fires, so a smaller subset of these has been used at NASA in a process called "fleet-leader" logic, in which successful operation of a small sample set of engines with certain criteria on starts, total test operation, and even estimate of total fatigue life are used to determine if a proposed engine is acceptable for flight. The sample set has been seen by historical experience to weed out the failures due to "infant mortality" (i.e., design failures that turn up immediately), and with accompanying analysis have been shown to be a valid compromise allowing engine verification at a reasonable cost.

Finally, hot-fire tests will identify failures that structural dynamic response can potentially play a role in. One example is high turbopump rotor response. Although this is generally a rotordynamic issue (a closely-related, but separate discipline from SD), an accurate SD model of the turbopump housing, such as the one described above, which provides the boundary conditions for the outer race of the turbine bearings often proves necessary.

## 7.4 Qualification and Acceptance Testing

This testing procedure is similar to modal testing and so is often confused with it, but its purpose is significantly different. A "qual" test is used generally to validate the successful operation of small components within a rocket engine or launch vehicle that are generally not analyzed with a finite element analysis, such as temperature or speed probes. A single sample of the component is bolted rigidly to a shaker table which then shakes it in three axes at wide-band spectral levels identified by conservative enveloping of flight vibration environments, as described in Sect. 5.4. The component is then examined for functionality both during the test and afterwards. Acceptance testing is also a type of shaker test, but is used to screen out unacceptable workmanship. It is very similar to qualification testing except is performed on every sample and generally at a level 6 dB lower [2]. Both of these types of testing fall into the "Shock and Vibro-Acoustics" discipline, to be discussed in Chap. 8.

## 7.5 Whirligig Testing

In conclusion, I will mention a specific type of development test called "whirligig" or "spin pit" testing. This test places an instrumented turbine wheel in a facility where it can be spun at close to operational speeds. The facility also usually has near-vacuum capability to enable inlet air jets or speakers placed at optimal locations around the periphery to provide traveling wave excitation like that inside the turbine [3]. The dynamic response of the blades can be measured with strain gages connected to a slip ring to get the signal out of the facility, or with laser tip-timing techniques, such as the Agilis$^©$ system. The spin pit is generally used to test blade damping techniques; since these dampers are usually highly nonlinear, analysis is quite difficult, so a dedicated test program is required to iteratively design and validate an acceptable damper. As these facilities are quite flexible, they can also be used for other turbine bladed-disk dynamic studies.

## References

1. Keenner C, D'Souza K, Design and Dynamic Characterization of the OSU Rotor 67 for Full Scale Damping and Mistuning Studies, ASME 2022 Turbomachinery Technical Conference & Exposition.
2. Ferebee R., Development of Vibroacoustic and Shock Design and Test Criteria Standard, MSFC-STD-3676, Rev. B, March 15, 2017
3. Li, Jia, Experimental investigation of mistuned bladed disks system vibration. Engineering Sciences [physics]. University of Michigan-Ann Arbor, 2007. English. fftel-00923790f.

# Related Disciplines and Advanced Topics     8

## 8.1    Related Disciplines

There are a number of closely-related disciplines to SD that the SD practitioner should be aware of and strive to gain a basic understanding of. I will not attempt to provide this understanding in this text but will just provide an introduction for awareness.

Structural Dynamics cannot be practiced in a vacuum (only figuratively) and a basic understanding of the other subjects covered in a standard mechanical, civil, or aerospace engineering curriculum is necessary. Strength and stress analysis, in particular fatigue analysis, is required; understanding high cycle fatigue is so critical that an overview is provided in Sect. 3.3. Of course, a basic understanding of linear and nonlinear stress versus strain, stress components and failure theories, ultimate strength, yield strength, fracture analysis, and stress concentrations is also necessary. Fluid dynamics is also important, particularly for turbomachinery and nozzles. A grasp of inviscid and viscous flows, the potential concept, subsonic versus supersonic flow, and Bernoulli's pressure, density, and velocity relationships is helpful. Acoustic relationships are also becoming more important, particularly for pump components operating in liquid propellants. Finally, thermal conductivity and gradients, expansion stresses, and thermoelastic materials also play a role in SD analysis.

There are three other related disciplines that will generally not be covered in an undergraduate curriculum. The first of these is rotordynamics, which can be thought of as structural dynamics on steroids in its difficulty. This discipline focuses on the stability of turbopump shafts during operation. The shaft (also called the rotor) can oscillate out of its nominal center-line location not only due to its natural frequencies, which are a function of heavily nonlinear stiffness of the bearings, but also due to unavoidable slight eccentricity of its center of mass. These effects cause a phenomenon called whirl, in which the center of mass rotates about the center-line at a rate different that the rotational rate

© The Author(s), under exclusive license to Springer Nature Switzerland AG 2022     163
A. M. Brown, *Structural Dynamics of Liquid Rocket Engines*, Synthesis Lectures on Mechanical Engineering, https://doi.org/10.1007/978-3-031-18207-5_8

of the shaft. This whirl must be kept stable, but a variety of influences can cause it to go unstable, including insufficient damping and "cross-coupled" stiffness and damping, which are forces perpendicular to the whirl deflection. Every component along the rotor will affect these quantities. The resulting instability can be catastrophic, as the bearings will be crushed, causing destruction of the bearing or contact between the rotor and housing, which can cause a cascade of engine system effects and even an explosion of the engine.

Another important related discipline is dynamic data analysis. This focuses on extracting meaningful measurements from test data. The difficulties in accomplishing this task are not obvious to someone not regularly dealing with the field, but are substantial, as completely different values of a quantity of interest (e.g., strain) can be obtained based purely on the processing parameters used. Signal processing theory is a major part of the discipline, in particular the study of sampling rates and Nyquist criteria, signal block size, and filtering. Random vibration and statistics concepts are also essential, including coherence of different signals.

Vibro-acoustics is the other graduate-level related discipline. This field focuses on the interaction of sound waves in a fluid with vibration in a solid. This interaction can be vibration induced noise, such as music speakers, noise causing vibration, including rocket launch noise impinging and damaging on payloads in the cargo bays, and sound transmitted through a structure and retransmitted on the other side [1]. This is particularly a problem for large panels of launch vehicles, which includes the nozzle for LRE's. This field uses an entirely different type of analytical technique, called Statistical Energy Analysis, rather than FEA, because the modal density is so high. As mentioned in the last chapter, this type of loading includes what is called "shock", which consists of very short duration impulsive events such as blast loads.

## 8.2    Advanced Topics

It should be clear at this point that structural dynamic evaluation of liquid rocket engine is an extremely complex, rich discipline, and there is a substantial field of published literature to reflect this. Much of this literature has resulted in techniques discussed in this text, but there is much more outside of the scope achievable in a single volume. A few of these topics will be mentioned in this section.

Mistuning is continuing to be one of the main areas of advanced study. Although the procedures presented in Sect. 3.6.7 allow initial evaluation of both large populations of bladed-disks as well as individual samples, there continue to be improvements in these methods as well as studies to bridge the gaps in our knowledge. The MISER FMM code, for instance, which enables evaluation of large sample populations necessary for design, has been significantly improved recently by incorporating factors relating the mode of interest to nearby modes, which will influence the MAF [2]. Another aspect of improvement is that almost all the techniques assume that the entire blade-to-blade variation is due to stiffness only, which can therefore be accounted for by varying Young's Modulus. In reality, the geometry of the blades is the random variable, so a number of researchers are examining how to incorporate this "geometric mistuning" into both the nondeterministic (i.e., large population) and individual sample dynamic evaluation process [3]. Another area of improvement is gaining a better understanding of the physics of mistuning localization and amplification through new tools such as the "band" diagram, which describes how the mistuned natural frequencies propagate around a nominal frequency [4].

A second advanced topic not only in SD of LRE's but structural mechanics in general, and for which mistuning is a subtopic, is structural dynamic analysis and uncertainty quantification of non-deterministic structures. Virtually all the components we have discussed in this text are in reality non-deterministic, i.e., both their geometric and material properties lie within a non-discrete, statistical distribution, resulting in natural frequencies that are also non-deterministic. This uncertainty is generally accounted for by using frequency margins between excitation and natural frequencies. Another way to explicitly identify this uncertainty is to use a band instead of a line for the horizontal natural frequencies in Campbell diagrams (this is much more difficult to do in interference diagrams because of the density of the modes on the plot, but the uncertainty should be recognized). Identifying an accurate natural frequency uncertainty band becomes more complicated with pump structures because of the large number of complicating factors, as described in Sect. 3.6.8. Traditionally, modal test is used to reduce the uncertainty, but actual quantification of how test should be used has not been extensively investigated; a new method has been identified using Conditional Covariance of the model frequencies of interest to the modal tested frequencies as a technique to provide a value of variance for a statistical distribution [5]. In essence, an estimate of the variance of the desired natural frequency using very conservative estimates of the variances of the random variables can be reduced significantly because specific values of some of these values are known from test; this is illustrated in Fig. 8.1 for two dimensions but is applicable for dimensions equal to the number of tested natural frequencies.

For forced response, where the desired output is the probability of failure, the probabilistic problem becomes even more complicated, as uncertainties in all the parameters must be considered, including loading, damping, excitation frequency, and natural frequency. There has been a good deal of development of methodologies using measured failures in engine systems to generate a "probabilistic risk assessment", which is called a "top-down" analysis. There has been less work trying to generate the probability of failure using the physics-based forced response parameters, though, which is called "bottoms-up", but several researchers have published studies in this field [6–8]. In any non-deterministic evaluation, it is critical to identify which uncertainties are "aleatory", or inherent in the component and irreducible, and which are "epistemic" and due to a lack

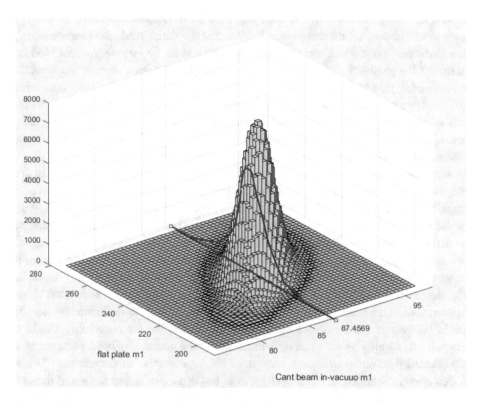

**Fig. 8.1** Multi-variate normal probability density function with 2-D slice at given information value

of knowledge or modeling capability, and which can be reduced by more accurate modeling, additional testing, etc. This categorization will assist significantly in the development of physics-based techniques.

The last advanced topic I will mention here is fluid/structural dynamic interaction in rocket engines. As discussed in Chap. 3, one-way application of CFD loads onto turbomachinery components has been practical for two decades now, although there are still situations when even this is cutting edge, such as pump-side forced response in cavitation, or loading within acoustically important environments. In addition, many situations now call for two-way coupled analysis, which is still a discipline under research and development. One example where this methodology becomes necessary is when trying to quantify aero or hydroelastic damping, where the structural stiffness and damping are functions of the flow velocity. Two-way analysis is also necessary for the evaluation of bellows in ducts and temperature probes because of the vortex shedding phenomena. It is also critical for evaluating the response of rocket nozzles due to the side load phenomenon, where the structural deformation is large enough to change the flow. Other than the calculation of aerodamping in turbines, which has and continues to be studied thoroughly, this field has had very little attention paid to it in SD of LRE's, with nothing in the literature beyond what has already been referenced. Therefore, this area is ripe for research and development, and can provide motivation for continued study and application of new methods in the exciting field of Structural Dynamics of Liquid Rocket Engines.

## References

1. Vibroacoustics of Aerospace Structures, Short Course at NASA/MSFC, ATA Engineering, October 1–4, 2007.
2. Hegde SS, Multi-row Aeromechanical and Aeroelastic Aspects of Embedded Gas Turbine Compressor Rotors, Ph.D. Dissertation, Duke University, December 2021.
3. Henry EB, Brown JM, Beck JA, Reduced Order Geometric Mistuning Models Using Principal Component Analysis Approximations, AIAA SciTech Forum, San Diego CA, 4–8 January 2016.
4. Rodriguez AM, Kauffman JL, Vibration Localization in Cyclic Structures: A Discrete Low-Order Model," Proc. AIAA Science and Technology Forum 2018, AIAA-2018-0183, January 8–12, Kissimmee, FL.
5. Brown, Andrew M., DeLessio, Jennifer L., Wray, Timothy J., Uncertainty Quantification of Inducer Eigenvalues using Conditional Assessment of Models and Modal Test of Simpler Systems, IMAC-XXXVIX Virtual Conference on Structural Dynamics, February 8–11, 2021, paper 10505.
6. Brown, Andrew M., DeHaye, Michael, DeLessio, Steven, Probabilistic Methods to Determine Resonance Risk and Appropriate Damping for Rocket Engine Turbine Blades, AIAA Journal of Propulsion and Power (2013). https://doi.org/10.2514/1.B34834.
7. Zhou K, Liang G, Tang J, Vibration analysis of structure with uncertainty using two-level Gaussian processes and Bayesian inference, J. Phys: Conf Series 744 012202, 2016.
8. Mullins, J., Mahadevan, S., Bayesian Uncertainty Integration for Model Calibration, Validation, and Prediction, Journal of Verification, Validation, and Uncertainty Quantification, March 2016, 1(1): 011006 (10 pages).

Printed in the United States
by Baker & Taylor Publisher Services